TECH/GENERAL RADIO AMATEUR FCC TEST MANUAL

by

MARTIN SCHWARTZ

**Published by
AMECO PUBLISHING CORP.**
220 E. JERICHO TURNPIKE
MINEOLA, NEW YORK 11501

TECH/GENERAL RADIO AMATEUR FCC TEST MANUAL

Copyright 1987
by the
Ameco Publishing Corp.

All rights reserved. This book, or parts thereof, may not be reproduced in any form, without the permission of the publisher.

Library of Congress Cat. No. 87-71509
ISBN No. 0-912146-25-7

Printed in the United States of America

PREFACE

This Tech/General FCC Test Manual is part of a series of books published by the Ameco Publishing Corp. for the purpose of preparing individuals for the Federal Communications Commission Amateur Radio Operator examinations. The first half of this book is for the Technician Class license examination. The second half is for the General Class license examination.

The questions and multiple choice answers in this manual have been issued by the Volunteer Examiner Coordinator's Committee under the supervision of the FCC. The 25 questions and multiple choice answers on the actual examination will be drawn from these questions. The questions in this book are divided into subelements. A specific number of questions (shown under each subelement heading), will be taken from each subelement. The discussions that follow the multiple choice answers were written by the author to give the reader sufficient background material to understand the reasons for choosing the proper multiple choice answer.

In those instances where this author feels that the correct multiple choice answer is complete and adequate for the proper understanding of the subject matter, there is no discussion; only the correct answer is indicated. In most of the questions, the discussions explain the correct multiple choice answers and give additional useful material that helps with the understanding of the questions and answers.

There are a few questions and/or multiple choice answers that are somewhat ambiguous. Since this author is not responsible for the questions or the multiple choice answers, they cannot be changed. However, in these few instances, the designated answer is given, together with a thorough explanation of the subject matter involved.

The FCC examination consists of 25 questions for the Technician Class license and 25 questions for the General Class license. 75% is the passing grade for each examination. Most VEC's require 19 correct answers.

Although this guide deals with Elements 3A and 3B of the Amateur Radio Operator's examinations, the Ameco Publishing Corp. publishes guides covering all the other amateur elements, as well as code courses for learning the International Morse Code. If additional theory background information is required, it is suggested that the Amateur Radio Theory Course, Cat. #102-01, be consulted. (See back cover for details). GOOD LUCK!

Martin Schwartz

TABLE OF CONTENTS

PREFACE 3

TECHNICIAN CLASS TEST MANUAL
    A. Rules and Regulations TA-1
    B. Operating Procedures TB-1
    C. Radio Wave Propagation TC-1
    D. Amateur Radio Practice TD-1
    E. Electrical Principles TE-1
    F. Circuit Components TF-1
    G. Practical Circuits TG-1
    H. Signals and Emissions TH-1
    I. Antennas and Feedlines TI-1

GENERAL CLASS TEST MANUAL
    A. Rules and Regulations GA-1
    B. Operating Procedures GB-1
    C. Radio Wave Propagation GC-1
    D. Amateur Radio Practice GD-1
    E. Electrical Principles GE-1
    F. Circuit Components GF-1
    G. Practical Circuits GG-1
    H. Signals and Emissions GH-1
    I. Antennas and Feedlines GI-1

RST REPORTING SYSTEM AP-1

FREQUENCY CHART AP-2

TABLE OF EMISSIONS AP-3

SUBELEMENT 3AA
RULES AND REGULATIONS
(5 questions)

3AA-1.1 What is the control point of an amateur station?
A. The operating position of an amateur radio station where the control operator function is performed
B. The operating position of any amateur radio station operating as a repeater user station
C. The physical location of any amateur radio transmitter, even if it is operated by radio link from some other location
D. The variable frequency oscillator (VFO) of the transmitter
The answer is A. Every amateur radio station must have at least one control point.

3AA-1.2 What is the term for the operating position of an amateur station where the control operator function is performed?
A. The operating desk   B. The control point
C. The station location   D. The manual control location
The answer is B. See answer 3AA-1.1.

3AA-2.1 What is an amateur emergency communication?
A. A nondirected request for help or a distress signal relating to the immediate safety of human life or the immediate protection of property
B. A communication with the manufacturer of the Amateur's equipment in case of equipment failure
C. The only type of communication allowed in the Amateur Radio Service
D. A communication that must be left to the Public Safety Radio Services; e.g., police and fire officials
The answer is A.

3AA-2.2 What is the term for an amateur radiocommunication directly related to the immediate safety of life of an individual?
A. Immediate safety communication B. Emergency communication
C. Third party communication   D. Individual communication
The answer is B. See answer 3AA-2.1.

3AA-2.3 What is the term for an amateur radiocommunication directly related to the immediate protection of property?
A. Emergency communication   B. Immediate communication
C. Property communication   D. Priority traffic
The answer is A. See answer 3AA-2.1.

3AA-2.4 Under what circumstances does the FCC declare that a general state of communications emergency exists?
A. When a declaration of war is received from Congress
B. When the maximum usable frequency goes above 28 MHz
C. When communications facilities in Washington, DC, are disrupted
D. In the event of an emergency disrupting normally available communication facilities in any widespread area(s)
The answer is D. In the event of an emergency, disrupting normally

available communications facilities in any widespread area, the Federal Communications Commission may declare that a general state of communications emergency exists. It designates the areas concerned, and specifies the amateur frequency bands, or segments of such bands, for use only by amateurs participating in emergency communication within or with such affected areas.

3AA-2.5 How does an amateur operator request the FCC to declare that a general state of communications emergency exists?
A. Communication with the FCC engineer-in-charge of the affected area
B. Communication with the U.S. senator or congressman for the area affected
C. Communication with the local Emergency Coordinator
D. Communication with the Chief of the FCC Private Radio Bureau

The answer is A. Amateurs desiring to request the declaration of such a state of emergency should communicate with the Commission's Engineer in Charge of the areas concerned. Whenever such declaration has been made, operation of and with amateur stations in the area concerned shall be only in accordance with the requirements set forth in this section, but such requirements shall in nowise affect other normal amateur communication in the affected area when conducted on frequencies not designated for emergency operation.

3AA-2.6 What type of instructions are included in an FCC declaration of a general state of communications emergency?
A. Designation of the areas affected and of organizations authorized to use radiocommunications in the affected area.
B. Designation of amateur frequency bands for use only by amateurs participating in emergency communications in the affected area, and complete suspension of Novice operating privileges for the duration of the emergency.
C. Designation of the areas affected and specification of the amateur frequency bands or segments of such bands for use only by amateurs participating in emergency communication within or with such affected area(s)
D. Suspension of amateur rules regarding station identification and business communication.

The answer is C.

3AA-2.7 What should be done by the control operator of an amateur station which has been designated by the FCC to assist in making known information relating to a general state of communications emergency?
A. The designated station shall act as an official liaison station with local news media and law-enforcement officials
B. The designated station shall monitor the designated emergency communications frequencies and warn noncomplying stations of the state of emergency
C. The designated station shall broadcast hourly bulletins from the FCC concerning the disaster situation
D. The designated station shall coordinate the operation of all phone-patch traffic out of the designated area

The answer is B. These control operators should monitor the designated amateur emergency communications bands, and warn noncomplying stations observed to be operating in those bands. Such station, when so designated, may transmit for that purpose on any frequency or frequencies authorized to be used by that station, provided such transmissions do not interfere with essential emergency communications in progress; however, such transmissions shall preferably be made on authorized frequencies immediately adjacent to those segments of the amateur bands being cleared for the emergency. Individual transmissions for the purpose of advising other stations of the existence of the communications emergency shall refer to this section by number (97.107) and shall specify, briefly and concisely, the date of the Commission's declaration, the area and nature of the emergency, and the amateur frequency bands or segments of such bands which constitute the amateur emergency communications bands at the time.

3AA-2.8 During an FCC-declared general state of communications emergency, how must the operation by, and with, amateur stations in the area concerned be conducted?
A. All transmissions within all designated amateur communications bands other than communications relating directly to relief work, emergency service, or the establishment and maintenance of efficient Amateur Radio networks for the handling of such communications shall be suspended
B. Operations shall be governed by part 97.93 of the FCC rules pertaining to emergency communications
C. No amateur operation is permitted in the area during the duration of the declared emergency
D. Operation by and with amateur stations in the area concerned shall be conducted in the manner the amateur concerned believes most effective to the speedy resolution of the emergency situation

The answer is A. Whenever such declaration has been made, operation of and with amateur stations in the area concerned shall be only in accordance with the requirements set forth in this section, but such requirements shall in nowise affect other normal amateur communication in the affected area when conducted on frequencies not designated for emergency operation.

3AA-3.1 Notwithstanding the numerical limitations in the FCC Rules, how much transmitting power shall be used by an amateur station?
A. There is no regulation other than the numerical limits
B. The minimum power level required to achieve S9 signal reports
C. The minimum power necessary to carry out the desired communication
D. The maximum power available as long as it is under the allowable limit

The answer is C.

3AA-3.6 What is the maximum transmitting power permitted an amateur station in beacon operation?
A. 10 watts PEP output
B. 100 watts PEP output
C. 500 watts PEP output
D. 1500 watts PEP output

The answer is B. The maximum PEP output in beacon operation shall not exceed 100 watts. However, these stations should use the minimum transmitting power necessary to carry out the desired communications.

3AA-3.8 What is the maximum transmitting power permitted an amateur station on 146.52-MHz?
A. 200 watts PEP output
B. 500 watts ERP
C. 1000 watts dc input
D. 1500 watts PEP output

The answer is D. 146.52 MHz is in the 2 meter band where the maximum power permitted is 1500 watts PEP output.

3AA-4.2 How must a newly-upgraded Technician control operator with a Certificate of Successful Completion of Examination identify the station while it is transmitting on 146.34-MHz pending receipt of a new operator license?
A. The new Technician may not operate on 146.34-MHz until his/her new license arrives
B. The licensee gives his/her call sign, followed by the word "temporary" and the identifier code shown on the certificate of successful completion
C. No special form of identification is needed
D. The licensee gives his/her call sign and states the location of the VE examination where he or she obtained the certificate of successful completion

The answer is B.

3AA-4.4 Which language(s) must be used when making the station identification by telephony?
A. The language being used for the contact may be used if it is not English, providing the U.S. has a third-party-traffic agreement with that country.
B. English must be used for identification.
C. Any language may be used, if the country which uses that language is a member of the International Telecommunication Union
D. The language being used for the contact must be used for identification purposes.

The answer is B.

3AA-4.5 What aid does the FCC recommend to assist in station identification when using telephony?
A. A speech-compressor
B. Q signals
C. An internationally recognized phonetic alphabet
D. Distinctive phonetics, made up by the operator and easy to remember

The answer is C. The Commission encourages the use of a nationally or internationally recognized standard phonetic alphabet as an aid for correct telephony identification.

3AA-4.6 What emission mode may always be used for station identification, regardless of the transmitting frequency?
A. A1A
B. F1B
C. A2B
D. A3E

The answer is A. Type A1A (telegraphy) may be used because it is permissible on all amateur frequencies.

3AA-5.1 Under what circumstances, if any, may a third-party participate in radiocommunications from an amateur station?
A. A control operator must be present and continuously monitor and supervise the radio communication to ensure compliance with the

rules. In addition, contacts may only be made with amateurs in the US and countries with which the US has a third-party traffic agreement
B. A control operator must be present and continuously monitor and supervise the radio communication to ensure compliance with the rules only if contacts are made with amateurs in countries with which the US has no third-party traffic agreement
C. A control operator must be present and continuously monitor and supervise the radio communication to ensure compliance with the rules. In addition, the control operator must key the transmitter and make the station identification.
D. A control operator must be present and continuously monitor and supervise the radio communication to ensure compliance with the rules. In addition, if contacts are made on frequencies below 30 MHz, the control operator must transmit the call signs of both stations involved in the contact at 10-minute intervals
The answer is A.

3AA-5.2 Where must the control operator be situated when a third-party is participating in radiocommunications from an amateur station?
A. If a radio remote control is used, the control operator may be physically separated from the control point, when provisions are incorporated to shut off the transmitter by remote control
B. If the control operator supervises the third party until he or she is satisfied of the competence of the third party, the control operator may leave the control point
C. The control operator must stay at the control point for the entire time the third party is participating
D. If the third party holds a valid radiotelegraph license issued by the FCC, no supervision is necessary
The answer is C.

3AA-5.3 What must the control operator do while a third-party is participating in radiocommunications?
A. If the third party holds a valid commercial radiotelegraph license, no supervision is necessary
B. The control operator must tune up and down 5 kHz from the transmitting frequency on another receiver, to insure that no interference is taking place
C. If a radio control link is available, the control operator may leave the room
D. The control operator must continuously monitor and supervise the radiocommunication to insure compliance with the rules
The answer is D.

3AA-5.4 Under what circumstances, if any, may a third party assume the duties of the control operator of an amateur station?
A. If the third party holds a valid commercial radiotelegraph license, he or she may act as control operator
B. Under no circumstances may a third party assume the duties of control operator
C. During Field Day, the third party may act as control operator
D. An Amateur Extra Class licensee may designate a third party as control

operator, if the station is operated above 450 MHz

The given answer is B. However, a third party MAY assume the functions of the control operator PROVIDED that he/she is a licensed radio amateur and is designated by the licensee of the station to be the control operator. The control operator may only operate the station to the extent permitted by his/her operator license.

3AA-6.3 What types of material compensation, if any, may be involved in third-party traffic transmitted by an amateur station?
A. Payment of an amount agreed upon by the amateur operator and the parties involved
B. Assistance in maintenance of auxiliary station equipment
C. Donation of amateur equipment to the control operator
D. No compensation may be accepted

The answer is D. Third party traffic involving material compensation, either tangible or intangible, direct or indirect, is forbidden.

3AA-6.4 What types of business communications, if any, may be transmitted by an amateur station on behalf of a third party?
A. Section 97.57 specifically prohibits business communications in the Amateur Service
B. Business communications involving the sale of amateur radio equipment
C. Business communications involving an emergency, as defined in Part 97
D. Business communications aiding a broadcast station

The answer is C.

3AA-6.5 When are third-party messages limited to those of a technical nature relating to tests, and to remarks of a personal character for which, by reason of their unimportance, recourse to the public telecommunications service is not justified?
A. Only when communicating with a person in a country with which the US does not share a third-party traffic agreement
B. When communicating with a non-profit organization such as the ARRL
C. When communicating with the FCC
D. Communications between amateurs in different countries are always limited to those of a technical nature relating to tests and remarks of a personal nature for which, by reason of their unimportance, recourse to the public telecommunications service is not justified

The answer is D. The limitations apply to transmissions between amateur stations of different countries. The question is actually a rephrasing of a paragraph in Article 32 of the International Radio Regulations of the International Telecommunications Union (ITU).

3AA-7.1 What kinds of one-way communications by amateur stations are not considered broadcasting?
A. All types of one-way communications by amateurs are considered by the FCC as broadcasting
B. Beacon operation, radio-control operation, emergency communications, information bulletins consisting solely of subject matter relating to amateur radio, roundtable discussions and code practice transmissions
C. Only code-practice transmissions conducted simultaneously on all available amateur bands below 30 MHz and conducted for more than 40

hours per week are not considered broadcasting
D. Only actual emergency communications during a declared communications emergency are exempt
The answer is B.

3AA-7.2 What is a one-way radiocommunication?
A. A communication in which propagation at the frequency in use supports signal travel in only one direction.
B. A communication in which different emissions are used in each direction
C. A communication in which an amateur station transmits to and receives from a station in a radio service other than amateur
D. A transmission to which no on-the-air response is desired or expected
The answer is D. One way radiocommunication is the transmission of information or signals from a transmitter without the expectation of a reply and without the back and forth conversation that one has with two-way transmissions.

3AA-7.3 What kinds of one-way information bulletins may be transmitted by amateur stations?
A. NOAA weather bulletins
B. Commuter traffic reports from local radio stations
C. Regularly scheduled announcements concerning Amateur Radio equipment for sale or trade
D. Bulletins consisting solely of information relating to amateur radio
The answer is D.

3AA-7.4 What types of one-way amateur radiocommunications may be transmitted by an amateur station?
A. Beacon operation, radio control, code practice, retransmission of other services
B. Beacon operation, radio control, transmitting an unmodulated carrier, NOAA weather bulletins
C. Beacon operation, radio control, information bulletins consisting solely of information relating to amateur radio, code practice and emergency communications transmissions
D. Beacon operation, emergency-drill-practice transmissions, automatic retransmission of NOAA weather transmissions, code practice
The answer is C. In addition to the above, round table discussions or net-type operations are permitted.

3AA-8.1 What are the HF privileges authorized to a Technician control operator?
A. 3700 to 3750 kHz, 7100 to 7150 kHz (7050 to 7075 kHz when terrestrial station location is in Alaska or Hawaii or outside Region 2), 14,100 to 14,150 kHz, 21,100 to 21,150 kHz, and 28,100 to 28,150 kHz only
B. 3700 to 3750 kHz, 7100 to 7150 kHz (7050 to 7075 kHz when terrestrial station location is in Alaska or Hawaii or outside Region 2), 21,100 to 21,200 kHz, and 28,100 to 28,500 kHz only
C. 28,000 to 29,700 kHz only
D. 3700 to 3750 kHz, 7100 to 7150 kHz (7050 to 7075 kHz when terrestrial station location is in Alaska or Hawaii or outside Region 2), and 21,100

to 21,200 kHz only

The answer is B. A Technician Class operator does not have any authorized frequency privileges in the High Frequency (1.8 to 30 MHz) amateur bands except those authorized to Novice Class operators.

3AA-8.2 Which operator licenses authorize privileges on 52.525MHz?
A. Extra, Advanced only
B. Extra, Advanced, General only
C. Extra, Advanced, General, Technician only
D. Extra, Advanced, General, Technician, Novice

The answer is C. 52.525 MHz is in the 6 meter band, a band in which all amateurs, except Novice Class operators, have privileges.

3AA-8.3 Which operator licenses authorize privileges on 146.52-MHz?
A. Extra, Advanced, General, Technician, Novice
B. Extra, Advanced, General, Technician only
C. Extra, Advanced, General only   D. Extra, Advanced only

The answer is B. 146.52 MHz is in the 2 meter band, a band in which all amateurs, except Novice Class operators, have privileges.

3AA-8.4 Which operator licenses authorize privileges on 223.50-MHz?
A. Extra, Advanced, General, Technician, Novice
B. Extra, Advanced, General, Technician only
C. Extra, Advanced, General only   D. Extra, Advanced only

The answer is A.

3AA-8.5 Which operator licenses authorize privileges on 446.0-MHz?
A. Extra, Advanced, General, Technician, Novice
B. Extra, Advanced, General, Technician only
C. Extra, Advanced, General only   D. Extra, Advanced only

The answer is B. 446.0 MHz is in the 420-450 MHz band, a band in which all amateurs, except Novice Class operators, have privileges.

3AA-10.9 On what frequencies within the 6 meter band may emission F3E be transmitted?
A. 50.0-54.0 MHz only   B. 50.1-54.0 MHz only
C. 51.0-54.0 MHz only   D. 52.0-54.0 MHz only

The answer is B. Type F3E emission stands for frequency modulated telephony. It may be used in the 50.1 to 54.0 MHz part of the 6 meter band.

3AA-10.10 On what frequencies within the 2 meter band may emission F3F be transmitted?
A. 144.1-148.0 MHz only   B. 146.0-148.0 MHz only
C. 144.0-148.0 MHz only   D. 146.0-148.0 MHz only

The answer is A. F3F emission stands for frequency modulated television. It may be used in the 144.1-148.0 MHz part of the 2 meter band.

3AA-11.1 What is the nearest to the band edge that the transmitting frequency can be set?
A. 3 kHz for single sideband and 1 kHz for CW
B. 1 kHz for single sideband and 3 kHz for CW
C. 1.5 kHz for single sideband and 0.05 kHz for CW
D. As near as the operator desires, providing that no sideband, harmonic, or spurious emission (in excess of that legally permitted) falls outside

the band

The answer is D. He may operate as close as possible to the band edge as long as the ENTIRE signal (including sidebands, etc.) is within the band edge limits. For instance, if his sidebands extend 3 kHz to the side of his carrier, his carrier must be at least 3 kHz away from the band edge. In addition, consideration must be given to the accuracy of the frequency measuring equipment. If the signal generator or other frequency measuring device is accurate to within 5 kHz in the band being used, then the operator must stay an additional 5 kHz away from the band edge.

3AA-11.2 When selecting the transmitting frequency, what allowance should be made for sideband emissions resulting from keying or modulation?
A. The sidebands must be adjacent to the authorized amateur radio frequency band in use
B. The sidebands must be harmonically-related frequencies that fall outside of the amateur radio frequency band in use
C. The sidebands must be confined within the authorized amateur radio frequency band occupied by the carrier
D. The sidebands must fall outside of the amateur radio frequency band in use so as to prevent interference to other amateur radio stations
The answer is C. See answer 3AA-11.1.

3AA-12.1 What is the maximum mean output power an amateur station is permitted in order to operate under the special rules for radio control of remote model craft and vehicles?
A. One watt   B. One milliwatt   C. Two watts   D. Three watts
The answer is A.

3AA-12.2 What information must be indicated on the writing affixed to the transmitter in order to operate under the special rules for radio control of remote model craft and vehicles?
A. Station call sign
B. Station call sign and operating times
C. Station call sign and licensee's name and address
D. Station call sign, class of license, and operating times
The answer is C.

3AA-12.3 What are the station identification requirements for an amateur station operated under the special rules for radio control of remote model craft and vehicles?
A. Once every ten minutes, and at the beginning and end of each transmission
B. Once every ten minutes
C. At the beginning and end of each transmission
D. Station identification is not required
The answer is D. Station identification is not required for transmissions directed only to a remote model craft or vehicle.

3AA-12.4 Where must the writing indicating the station call sign and the licensee's name and address be affixed in order to operate under the special rules for radio control of remote model craft and vehicles?
A. It must be in the operator's possession

B. It must be affixed to the transmitter
C. It must be affixed to the craft or vehicle
D. It must be filed with the nearest FCC Field Office
  The answer is B. The writing should be on a durable material.

3AA-13.3 What is the maximum sending speed permitted for an emission F1B transmission between 28- and 50-MHz?
A. 56 kilobauds   B. 19.6 kilobauds   C. 1200 bauds   D. 300 bauds
  The answer is C. See discussion in question 3BA-13.2.

3AA-13.4 What is the maximum sending speed permitted for an emission F1B transmission between 50- and 220-MHz?
A. 56 kilobauds   B. 19.6 kilobauds   C. 1200 bauds   D. 300 bauds
  The answer is B. See discussion in question 3BA-13.2.

3AA-13.5 What is the maximum sending speed permitted for an emission F1B transmission above 220-MHz?
A. 300 bauds   B. 1200 bauds   C. 19.6 kilobauds   D. 56 kilobauds
  The answer is D.

3AA-13.6 What is the maximum frequency shift permitted for emission F1B when transmitted below 50-MHz?
A. 100 Hz   B. 500 Hz   C. 1000 Hz   D. 5000 Hz
  The answer is C. The frequency shift is the difference between the frequency for the "mark" signal and that for the "space" signal.

3AA-13.7 What is the maximum frequency shift permitted for emission F1B when transmitted above 50-MHz?
A. 100 Hz or the sending speed, in bauds, whichever is greater
B. 500 Hz or the sending speed, in bauds, whichever is greater
C. 1000 Hz or the sending speed, in bauds, whichever is greater
D. 5000 Hz or the sending speed, in bauds, whichever is greater
  The answer is C.

3AA-13.8 What is the maximum bandwidth permitted an amateur station transmission between 50- and 220-MHz using a non-standard digital code?
A. 20 kHz   B. 50 kHz   C. 80 kHz   D. 100 kHz
  The answer is A. In this context, the bandwidth is defined as the width of the frequency band, outside of which the mean power of any emission is attenuated by at least 26 decibels below the mean power of the total emission, a 3 kHz sampling bandwidth being used by the FCC in making this determination.

3AA-13.9 What is the maximum bandwidth permitted an amateur station transmission between 220- and 902-MHz using a non-standard digital code?
A. 20 kHz   B. 50 kHz   C. 80 kHz   D. 100 kHz
  The answer is D. See answer 3AA-13.8.

3AA-13.10 What is the maximum bandwidth permitted an amateur station transmission above 902-MHz using a non-standard digital code?
A. 20 kHz   B. 100 kHz
C. 200kHz, as defined by Section 97.66(g)
D. Any bandwidth, providing that the emission is in accordance with

sections 97.63(b) and 97.73(c)

The answer is D. On frequencies above 902 MHz, any bandwidth may be used provided the sideband frequencies resulting from keying or modulating a carrier wave shall be confined within the authorized amateur band and any spurious emissions or radiations are reduced or eliminated in accordance with good engineering practice.

3AA-14.1 What is meant by the term broadcasting?
A. The dissemination of radio communications intended to be received by the public directly or by intermediary relay stations
B. Retransmission by automatic means of programs or signals emanating from any class of station other than amateur
C. The transmission of any one-way radio communication, regardless of purpose or content
D. Any one-way or two-way radio communication involving more than two stations

The answer is A.

3AA-14.2 What classes of stations may be automatically retransmitted by an amateur station?
A. FCC licensed commercial stations
B. Federally or state-authorized Civil Defense stations
C. Amateur radio stations
D. National Weather Service bulletin stations

The answer is C. Repeaters may only be used to retransmit the radio signals of other amateur stations.

3AA-14.3 Under what circumstances, if any, may a broadcast station retransmit the signals from an amateur station?
A. Under no circumstances
B. When the amateur station is not used for any activity directly related to program production or newsgathering for broadcast purposes
C. If the station rebroadcasting the signal feels that such action would benefit the public
D. When no other forms of communication exist

The answer is B.

3AA-14.5 Under what circumstances, if any, may an amateur station retransmit a NOAA weather station broadcast?
A. If the NOAA broadcast is taped and retransmitted later
B. If a general state of communications emergency is declared by the FCC
C. If permission is granted by NOAA for amateur retransmission of the broadcast
D. Under no circumstances

The answer is D. NOAA stands for National Oceanic and Atmospheric Administration. Under no circumstances may an amateur retransmit any transmissions from any station other than an amateur station.

3AA-14.7 Under what circumstances, if any, may an amateur station be used for an activity related to program production or news-gathering for broadcast purposes?
A. The programs or news produced with the assistance of an amateur station must be taped for broadcast at a later time

B. An amateur station may be used for newsgathering and program production only by National Public Radio
C. Under no circumstances
D. Programs or news produced with the assistance of an amateur station must mention the call sign of that station

The answer is C. An amateur station may not be used for broadcasting. Nevertheless, an amateur may consent to a broadcast station retransmitting his transmissions and receptions. However, the broadcaster may not employ an amateur as it would a reporter.

3AA-15.2 Under what circumstances, if any, may singing be transmitted by an amateur station?
A. When the singing produces no dissonances or spurious emissions
B. When it is used to jam an illegal transmission
C. Only above 1215 MHz
D. Transmitting music is not permitted in the Amateur Service

The answer is D. Singing is considered music and the transmission of music by an amateur station is prohibited.

3AA-17.1 Under what circumstances, if any, may an amateur station transmit radiocommunications containing obscene words?
A. Obscene words are permitted when they do not cause interference to any other radio communication or signal
B. Obscene words are prohibited in amateur radio transmissions
C. Obscene words are permitted when they are not retransmitted through repeater or auxiliary stations
D. Obscene words are permitted, but there is an unwritten rule among amateurs that they should not be used on the air

The answer is B. An amateur may not transmit communications containing obscene, indecent or profane words, language or meaning.

3AA-17.2 Under what circumstances, if any, may an amateur station transmit radiocommunications containing indecent words?
A. Indecent words are permitted when they do not cause interference to any other radio communication or signal
B. Indecent words are permitted when they are not retransmitted through repeater or auxiliary stations
C. Indecent words are permitted, but there is an unwritten rule among amateurs that they should not be used on the air
D. Indecent words are prohibited in amateur radio transmissions

The answer is D. See answer to question 3AA-17.1

3AA-17.3 Under what circumstances, if any, may an amateur station transmit radiocommunications containing profane words?
A. Profane words are permitted when they are not retransmitted through repeater or auxiliary stations
B. Profane words are permitted, but there is an unwritten rule among amateurs that they should not be used on the air
C. Profane words are prohibited in amateur radio transmissions
D. Profane words are permitted when they do not cause interference to any other radio communication or signal

The answer is C. See answer to question 3AA-17.1

SUBELEMENT 3AB
OPERATING PROCEDURES
(3 questions)

3AB-1.1 What is the meaning of: "Your report is five-seven..."?
A. Your signal is perfectly readable and moderately strong
B. Your signal is perfectly readable, but weak
C. Your signal is readable with considerable difficulty
D. Your signal is perfectly readable with near pure tone

The answer is A. Radio amateur operators use the RST Reporting System in describing the quality of a signal to the amateur with whom they are communicating. The RST Reporting System is a means of rating the quality of a signal on a numerical basis. In this system, the first number stands for readability (R) and is rated on a scale of 1 to 5. The second number stands for signal strength (S) and is rated on a scale of 1 to 9. The third number (for CW operation) indicates the quality of the CW tone (T), and its scale is also 1 to 9. The higher the number, the better the signal. See page AP-1 for the complete RST Reporting System.
If we look at the Reporting System table on page AP-1, we see that "five" in the readability column stands for "perfectly readable" and "seven" in the signal strength column stands for "moderately strong".

3AB-1.2 What is the meaning of: "Your report is three-three"?
A. Your signal is readable with considerable difficulty and weak in strength
B. The station is located at latitude 33 degrees
C. The contact is serial number thirty-three
D. The contact is unreadable, very weak in strength

The answer is A. See answer 3AB-1.1. If we look at the Reporting System table on page AP-1, we see that the first number "three" indicates "readable with considerable difficulty". We also see that the second number "three" indicates a signal strength of "weak signals".

3AB-1.3 What is the meaning of: "Your report is five nine plus 20dB..."?
A. Your signal strength has increased by a factor of 100
B. Repeat your transmission on a frequency 20 kHz higher
C. The bandwidth of your signal is 20 decibels above linearity
D. A relative signal-strength meter reading is 20 decibels greater than strength 9

The answer is D. The five indicates that the readability is "perfectly readable". The nine indicates an extremely strong signal. The "plus dB" is part of the RST signal report given in conjunction with the signal strength. If the signal strength is greater than 9, then we can say that the signal strength is 9 plus 10dB or 9 plus 20 dB, etc.
Many communications receivers have an "S "meter built into them. The "S" meter indicates the relative signal strength of the received signal.
The lower half (left side) of the face of the S meter is calibrated in S units, from S-1 to S-9. Each S unit is approximately 5 to 6 dB. Depending upon the receiver, an S-9 reading indicates a signal at the antenna of

anywhere from 25 to 50 microvolts. The upper half of the meter is calibrated in dB over S-9, usually from 0 to 50 or 60 dB.

It is important to keep in mind the fact that the S meter is not an accurate signal strength meter, but rather a means of obtaining relative signal strength indications.

3AB-1.6 How should the microphone gain control be adjusted on an emission F3E transmitter?
A. For proper deviation on modulation peaks
B. For maximum, non-clipped amplitude on modulation peaks
C. For moderate movement of the ALC meter on modulation peaks
D. For a dip in plate current

The answer is A. F3E stands for frequency modulated telephony. By limiting the audio gain, we prevent excessive deviation with its subsequent interference to adjacent frequencies. On the other hand, the audio should be strong enough to provide reasonable deviation which produces a strong FM signal.

3AB-1.7 How is the call sign WE5TZD stated phonetically?
A. Whiskey-Echo-Foxtrot-Tango-Zulu-Delta
B. Washington-England-Five-Tokyo-Zanzibar-Denmark
C. Whiskey-Echo-Five-Tango-Zulu-Delta
D. Whiskey-Easy-Five-Tear-Zebra-Dog

The answer is C. If conditions are poor during telephony operation, or if you feel that you may not be understood, phonetics should be used. The International Telecommunication Union (ITU) recommends the following phonetic alphabet:

A - Alpha	G - Golf	N - November	U - Uniform
B - Bravo	H - Hotel	O - Oscar	V - Victor
C - Charlie	I - India	P - Papa	W - Whiskey
D - Delta	J - Juliette	Q - Quebec	X - Xray
E - Echo	K - Kilo	R - Romeo	Y - Yankee
F - Foxtrot	L - Lima	S - Sierra	Z - Zulu
	M - Mike	T - Tango	

3AB-1.8 How is the call sign KC4HRM stated phonetically?
A. Kilo-Charlie-Four-Hotel-Romeo-Mike
B. Kilowatt-Charlie-Four-Hotel-Roger-Mexico
C. Kentucky-Canada-Four-Honolulu-Radio-Mexico
D. King-Charlie-Foxtrot-Hotel-Roger-Mary

The answer is A. See answer 3AB-1.7.

3AB-1.9 How is the call sign AF6PSQ stated phonetically?
A. America-Florida-Six-Portugal-Spain-Quebec
B. Adam-Frank-Six-Peter-Sugar-Queen
C. Alpha-Fox-Sierra-Papa-Santiago-Queen
D. Alpha-Foxtrot-Six-Papa-Sierra-Quebec

The answer is D. See answer 3AB-1.7.

3AB-1.10 How is the call sign NB8LXG stated phonetically?
A. November-Bravo-Eight-Lima-Xray-Golf
B. Nancy-Baker-Eight-Love-Xray-George
C. Norway-Boston-Eight-London-Xray-Germany

OPERATING PROCEDURES

TB-3

D. November-Bravo-Eight-London-Xray-Germany
The answer is A. See answer 3AB-1.7.

3AB-1.11 How is the call sign KJ1UOI stated phonetically?
A. King-John-One-Uncle-Oboe-Ida
B. Kilowatt-George-India-Uncle-Oscar-India
C. Kilo-Juliette-One-Uniform-Oscar-India
D. Kentucky-Juliette-One-United-Ontario-Indiana
The answer is C. See answer to question 3AB-1.7.

3AB-1.12 How is the call sign WV2BPZ stated phonetically?
A. Whiskey-Victor-Two-Bravo-Papa-Zulu
B. Willie-Victor-Two-Baker-Papa-Zebra
C. Whiskey-Victor-Tango-Bravo-Papa-Zulu
D. Willie-Virginia-Two-Boston-Peter-Zanzibar
The answer is A. See answer to question 3AB-1.7.

3AB-1.13 How is the call sign NY3CTJ stated phonetically?
A. Norway-Yokohama-Three-California-Tokyo-Japan
B. Nancy-Yankee-Three-Cat-Texas-Jackrabbit
C. Norway-Yesterday-Three-Charlie-Texas-Juliette
D. November-Yankee-Three-Charlie-Tango-Juliette
The answer is D. See answer to question 3AB-1.7.

3AB-1.14 How is the call sign KG7DRV stated phonetically?
A. Kilo-Golf-Seven-Denver-Radio-Venezuela
B. Kilo-Golf-Seven-Delta-Romeo-Victor
C. King-John-Seven-Dog-Radio-Victor
D. Kilowatt-George-Seven-Delta-Romeo-Video
The answer is B. See answer to question 3AB-1.7.

3AB-1.15 How is the call sign WX9HKS stated phonetically?
A. Whiskey-Xray-Nine-Hotel-Kilo-Sierra
B. Willie-Xray-November-Hotel-King-Sierra
C. Washington-Xray-Nine-Honolulu-Kentucky-Santiago
D. Whiskey-Xray-Nine-Henry-King-Sugar
The answer is A. See answer to question 3AB-1.7.

3AB-1.16 How is the call sign AE0LQY stated phonetically?
A. Able-Easy-Zero-Lima-Quebec-Yankee
B. Arizona-Equador-Zero-London-Queen-Yesterday
C. Alfa-Echo-Zero-Lima-Quebec-Yankee
D. Able-Easy-Zero-Love-Queen-Yoke
The answer is C. See answer to question 3AB-1.7.

3AB-2.5 What is meant by the term AMTOR?
A. AMTOR is a system using two separate antennas with a common receiver to reduce transmission errors
B. AMTOR is a system in which the transmitter feeds two antennas, at right angles to each other, to reduce transmission errors
C. AMTOR is a system using independent sideband to reduce transmission errors
D. AMTOR is a system using error-detection and correction to reduce transmission errors

The answer is D. "AMTOR" stands for Amateur Teleprinting Over Radio. It is a code that is derived from the Baudot Code. AMTOR adds two data bits to the Baudot Code to give it seven data bits. In almost all the characters, one bit is added in front of the five Baudot bits and one bit is added after the five Baudot bits. For instance, the five data pulses of the letter "J" as shown in Figure 3AB-2.3A are 11010. In the AMTOR Code, we add a 1 in front and a 0 in back to give us 1110100 for the letter "J".

The 0's or 1's that are added to form the AMTOR Code always result in combinations of three or four 0's and three or four 1's. This limits the total number of combinations in the AMTOR Code to 35.

AMTOR uses a "time diversity" system to reduce transmission errors. By "time diversity", we mean that after the original signal is sent out it may be repeated some time later in order to correct for any errors. In one type of time diversity, each character is sent out twice. In a second type of time diversity the two stations are constantly checking each other and if one station has missed something it requests a repeat of the other station.

3AB-2.7 What is the most common frequency shift for emission F2B transmissions in the amateur VHF bands?
A. 85 Hz.   B. 170 Hz.   C. 300 Hz.   D. 425 Hz.

The answer is B. F2B is the emission designator for FM tone modulated telegraphy for automatic reception (RTTY).

3AB-2.8 What is an RTTY Mailbox?
A. A QSL Bureau for teletype DX cards
B. An open net for RTTY operators
C. An address to which RTTY operators may write for technical assistance
D. A system by which messages may be stored electronically for later retrieval

The answer is D. An RTTY Mailbox is an electronic "mailbox". Messages can be entered into the computer of the mailbox using a specific code. Messages can later be retrieved from the mailbox with the aid of the correct code or password.

3AB-2.9 What is the purpose of transmitting a string of RYRYRY characters in RTTY?
A. It is the RTTY equivalent of CQ
B. Since it represents alternative upper and lower case signals, it is used to assist the receiving operator in checking the shift mechanism
C. Since it contains alternating mark and space frequencies, it is a check on proper operation of the transmitting and receiving equipment
D. It is sent at the beginning of an important message to activate stations equipped with SELCAL and Autostart

The answer is C. In the Baudot code, the bit patterns for the letters "R" and "Y" are 01010 and 10101 respectively. These patterns are complementary and cause the most mechanical parts to move each character. This gives us a good check on the proper operation of the equipment.

3AB-3.1 How should a QSO be initiated through a station in repeater operation?
A. Say "breaker, breaker 79."

OPERATING PROCEDURES TB-5

B. Call the desired station and then identify your own station
C. Call "CQ" three times and identify three times
D. Wait for a "CQ" to be called and then answer it

The answer is B. A repeater is a receiver-transmitter device that receives a signal on one frequency and automatically retransmits it on another frequency. The repeater is located on a hill or other high point, and its purpose is to extend the range of communications of low power hand-held and mobile stations. The low power station transmits to a nearby repeater. The repeater then retransmits the signal over a much greater distance than the low power station could transmit.

If you wish to use a repeater, you should, with a brief call, make your presence known. Do not call CQ. If you wish to call someone, then call that station and give your own call.

3AB-3.2 Why should users of a station in repeater operation pause briefly between transmissions?
A. To check the SWR of the repeater
B. To reach for pencil and paper for third party traffic
C. To listen for any hams wanting to break in
D. To dial up the repeater's autopatch

The answer is C. Repeater transmissions should be kept short, with frequent pauses between transmissions for breakers. It allows those with emergencies or important information to break in.

3AB-3.3 Why should users of a station in repeater operation keep their transmissions short and thoughtful?
A. A long transmission may prevent someone with an emergency from using the repeater
B. To see if the receiving station operator is still awake
C. To give any non-hams that are listening a chance to respond
D. To keep long-distance charges down

The answer is A. See answer 3AB-3.2.

3AB-3.4 Why should simplex be used where possible instead of using a station in repeater operation?
A. Farther distances can be reached
B. To avoid long distance toll charges
C. To avoid tying up the repeater unnecessarily
D. To permit the testing of the effectiveness of your antenna

The answer is C. Simplex means "alternating transmissions between two or more stations using one frequency". It is communicating with another station directly instead of using a repeater. Simplex should always be used, where possible, instead of tying up a repeater.

3AB-3.5 What is the proper procedure to break into an on-going QSO through a station in repeater operation?
A. Wait for the end of a transmission and start calling
B. Shout, "break, break!" to show that you're eager to join the conversation
C. Turn on your 100-watt amplifier and override whoever is talking
D. During a break between transmissions, send your call sign

The answer is D. The term "break-break" should be used only when you have emergency communications.

3AB-3.6 What is the purpose of repeater operation?
A. To cut your power bill by using someone's higher power system
B. To enable mobile and low-power stations to extend their usable range
C. To reduce your telephone bill
D. To call the ham radio distributor 50 miles away
   The answer is B. See answer 3AB-3.1.

3AB-3.7 What is a repeater frequency coordinator?
A. Someone who coordinates the assembly of a repeater station
B. Someone who provides advice on what kind of system to buy
C. The club's repeater trustee
D. A person or group that recommends frequency pairs for repeater usage
   The answer is D. A frequency coordinator is a volunteer who recommends specific repeater frequencies so as to avoid interference with other repeaters.

3AB-3.9 What is the usual input/output frequency separation for stations in repeater operation in the 2-meter band?
A. 1 MHz   B. 1.6 MHz   C. 170 MHz   D. 0.6 MHz
   The answer is D. The separation of 0.6 MHz is in general use on 2 meters. This separation is not specified in the FCC Rules, but is agreed upon by the local users of the repeater.

3AB-3.10 What is the usual input/output frequency separation for stations in repeater operation in the 70 centimeter band?
A. 1.6 MHz   B. 5 MHz   C. 600 kHz   D. 5 kHz
   The answer is B. The 70 centimeter band is the 420-450 MHz band. See answer 3AB-3.9.

3AB-3.11 What is the usual input/output frequency separation for a 6 meter station in repeater operation?
A. 1 MHz   B. 600 kHz   C. 1.6 MHz   D. 20 kHz
   The answer is A. See answer 3AB-3.9.

3AB-3.13 What is the usual input/output frequency separation for a 1.25 meter station in repeater operation?
A. 1000 kHz   B. 600 kHz   C. 1600 kHz   D. 1.6 GHz
   The answer is C. 1.25 meters is the 220-225 MHz band. See answer 3AB-3.9.

3AB-6.4 Why should local amateur radiocommunications be conducted on VHF and UHF frequencies?
A. To minimize interference on HF bands capable of long-distance skywave communication
B. Because greater output power is permitted on VHF and UHF
C. Because HF transmissions are not propagated locally
D. Because absorption is greater at VHF and UHF frequencies
   The answer is A. The HF bands (3-30 MHz) are limited in frequency space and are generally crowded with stations doing DX (long distance) operating. It would be poor operating courtesy to use these bands for local communication and cause interference when a VHF band (30-300 MHz) could be used where there is much more frequency space available.

OPERATING PROCEDURES   TB-7

3AB-6.5 How can on-the-air transmissions be minimized during a lengthy transmitter testing or loading up procedure?
A. Use a dummy antenna
B. Choose an unoccupied frequency
C. Use a non-resonant antenna
D. Use a resonant antenna that requires no loading up procedure

The answer is A. A dummy antenna is a resistive device that places the same load on the transmitter that the antenna does. The dummy load does not radiate RF signals into space as the antenna does. Therefore, when testing or tuning the transmitter, use a dummy load. The final BRIEF "touch-up" tuning can be done on-the-air prior to actual operating.

3AB-6.6 When a frequency conflict arises between a simplex operation and a repeater operation, why does good amateur practice call for the simplex operation to move to another frequency?
A. The repeater's output power can be turned up to ruin the front end of the station in simplex operation
B. There are more repeaters than simplex operators
C. Changing the repeater's frequency is not practical
D. Changing a repeater frequency requires the authorization of the Federal Communications Commission

The answer is C. It is a simple matter for an amateur, operating simplex, to move his carrier frequency to another frequency. On the other hand, the repeater's frequencies are listed, known and used by many operators, and it would be highly impractical to change them.

3AB-6.7 What should be done before installing an amateur station within one mile of an FCC monitoring station?
A. The amateur should apply to the FCC for a Special Temporary Authority for operation within the shadow of the monitoring facility's antenna system
B. The amateur should make sure a line-of-sight path does not exist between the amateur station and the monitoring facility
C. The amateur should consult with the Commission to protect the monitoring facility from harmful interference
D. The amateur should make sure the effective radiated power of the amateur station will be less than 200 watts PEP in the direction of the monitoring facility

The answer is C. It is important to consult with the FCC Field Operations Bureau prior to applying for a station license. The FCC may insert a clause in the amateur's license that protects the FCC from interference.

3AB-6.8 What is the proper Q signal to use to determine whether a frequency is in use before making a transmission?
A. QRL?   B. QRU?   C. QRV?   D. QRZ?

The answer is A. After sending QRL, listen before transmitting.

3AB-6.9 What is meant by "making the repeater time out"?
A. The repeater's battery supply has run out
B. The repeater's transmission time limit has expired during a single transmission
C. The warranty on the repeater duplexer has expired

D. The repeater is in need of repairs

The answer is B. FCC rules require that provisions must be incorporated to limit transmission to a period of no more than 3 minutes in the event of a malfunction in the control link. Most repeaters use a 3 minute timer that automatically shuts the repeater in 3 minutes if the control link fails. This 3 minute timer can also prevent an amateur from making lengthy transmissions without pausing. If he speaks more than 3 minutes continuously the repeater will "time-out". On the other hand, if he pauses frequently (which he should do) he will reset the timer and prevent it from "timing-out".

3AB-6.10 During commuting rush hours, which types of operation should relinquish the use of the repeater?
A. Mobile operators
B. Low-power stations
C. Highway traffic information nets
D. Third-party traffic nets

The answer is D. One of the primary uses of repeaters is as an aid to mobile operation. During commuter rush hours, mobile stations should have priority over base stations or third party traffic nets.

3AB-9.1 What is the proper distress calling procedure when using telephony?
A. Transmit MAYDAY
B. Transmit QRRR
C. Transmit QRZ
D. Transmit SOS

The answer is A.

3AB-9.2 What is the proper distress calling procedure when using telegraphy?
A. Transmit MAYDAY
B. Transmit QRRR
C. Transmit QRZ
D. Transmit SOS

The answer is D.

SUBELEMENT 3AC
RADIO WAVE PROPAGATION
(3 questions)

3AC-1.1 What is the ionosphere?
A. That part of the upper atmosphere where enough ions and free electrons exist to affect radio-wave propagation
B. The boundary between two air masses of different temperature and humidity, along which radio waves can travel
C. The ball that goes on top of a mobile whip antenna
D. That part of the atmosphere where weather takes place

The answer is A. The ionosphere is a gaseous region in the upper atmosphere that extends approximately 40 to 300 miles above the earth. It consists of several layers of ionized particles. The ionization is caused by the air particles being bombarded by the sun's ultraviolet rays and cosmic rays.

Figure 3AC-1.1 shows the makeup of the ionosphere during the daytime. Letters are assigned to the various gaseous layers of the ionosphere. The height above earth of each layer is also given. During the evening, the D and E layers disappear and the F2 and F1 layers combine to form a single F layer having a vertical height above earth of about 175 miles.

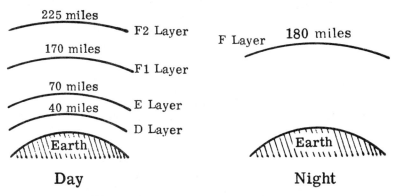

Day   Night

Figure 3AC-1.1. The ionosphere.

3AC-1.2 Which ionospheric layer limits daytime radiocommunications in the 80 meter band to short distances?
A. D layer   B. F1 layer   C. E layer   D. F2 layer

The answer is A. During the daytime, 80 meter signals are absorbed by the "D" layer and communication is limited to under 300 miles.

3AC-1.3 What is the region of the outer atmosphere which makes long-distance radiocommunications possible as a result of bending of radio waves?
A. Troposphere   B. Stratosphere   C. Magnetosphere   D. Ionosphere

The answer is D. See answers to questions 3AC-1.1 and 3AC-1.14.

3AC-1.4 Which layer of the ionosphere is mainly responsible for long-distance sky-wave radio communication?
A. D layer   B. E layer   C. F1 layer   D. F2 layer
The answer is D. During the daytime, F2 is responsible for long distance sky-wave communications. At nighttime, it is the F layer. See answer 3AC-1.1.

3AC-1.5 What are the two distinct sub-layers of the F layer of the ionosphere during the daytime?
A. Troposphere and stratosphere
B. F1 and F2
C. Electrostatic and electromagnetic
D. D and E
The answer is B. See answer 3AC-1.1.

3AC-1.8 What is the lowest region of the ionosphere that is useful for long-distance radio wave propagation?
A. The D layer   B. The E layer
C. The F1 layer   D. The F2 layer
The answer is B. The lowest layer is the D layer. However, it is more of an absorbing layer than a reflecting layer. It is therefore not useful for long distance communication. The next lowest layer is the E layer which IS useful for long distance communication.

3AC-1.11 What type of solar radiation is most responsible for ionization in the outer atmosphere?
A. Thermal   B. Ionized particle
C. Ultraviolet   D. Microwave
The answer is C. The ultraviolet radiation from the sun causes a large proportion of the air particles in the upper atmosphere to become ionized.

3AC-1.12 What is the lowest ionospheric layer?
A. The A layer   B. The D layer   C. The E layer   D. The F layer
The answer is B. See answer 3AC-1.1.

3AC-1.14 What is the region of the outer atmosphere which makes long-distance radiocommunications possible as a result of bending of the radio waves?
A. The ionosphere   B. The troposphere
C. The magnetosphere   D. The stratosphere
The answer is A. When a radio wave strikes the ionosphere, it is bent around or REFRACTED back to the earth some distance away from the transmitter. The ability of the ionosphere to return a radio wave back to earth depends upon the ion density, the frequency of the signal, the angle of radiation and other factors.
Most long distance communication results from the transmitted sky wave striking the "F" layer and returning to earth a considerable distance away. See Figure 3AC-1.14.

3AC-2.1 Which layer of the ionosphere is most responsible for absorption of radio signals during daylight hours?
A. The E layer   B. The F1 layer   C. The F2 layer   D. The D layer
The answer is D. See answers 3AC-1.2 and 3AC-1.8.

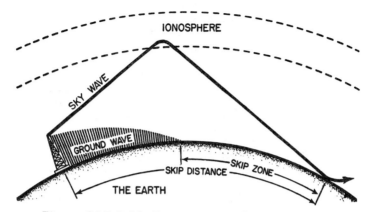

Figure 3AC-1.14. Propagation of radio waves.

3AC-2.2 When is ionospheric absorption most pronounced?
A. When radio waves enter the D layer at low angles
B. When tropospheric ducting occurs
C. When radio waves travel to the F layer
D. When a temperature inversion occurs

The answer is A. Ionospheric absorption is most pronounced during the daytime, especially at noontime.

3AC-2.5 During daylight hours, what effect does the D layer of the ionosphere have on 80 meter radio waves?
A. The D layer absorbs the signals
B. The D layer bends the radio waves out into space
C. The D layer refracts the radio waves back to earth
D. The D layer has little or no effect on 80 meter radio wave propagation

The answer is A. The D layer absorbs energy from 80 meter signals and limits their range to short distances. See answers 3AC-1.2 and 3AC-1.8.

3AC-2.6 What causes ionospheric absorption of radio waves?
A. A lack of D layer ionization
B. D layer ionization
C. The presence of ionized clouds in the E layer
D. Splitting of the F layer

The answer is B. The radiation from the sun causes ionization in the ionosphere. When the radio waves collide with the ionized particles, the radio waves give up energy in the form of heat. The lost energy or absorption is more pronounced at lower frequencies and during the daytime.

3AC-3.1 What is the highest radio frequency that will be refracted back to earth called?
A. Lowest usable frequency B. Optimum working frequency
C. Ultra high frequency   D. Critical frequency

The answer is D. When a signal is directed up to the ionosphere, it is refracted and returned to earth. As we gradually increase the frequency, we find that above a certain frequency, called the CRITICAL FREQUENCY, the signal is no longer returned to earth.

3AC-3.2 What causes the maximum usable frequency to vary?
A. Variations in the temperature of the air at ionospheric levels
B. Upper-atmospheric wind patterns
C. Presence of ducting
D. The amount of ultraviolet and other types of radiation received from the sun

The answer is D. Some of the factors that cause the maximum usable frequency to vary are: the time of day, the season, the degree of solar radiation, and the direction of the signal.

3AC-3.5 What does the term maximum usable frequency refer to?
A. The maximum frequency that allows a radio signal to reach its destination in a single hop
B. The minimum frequency that allows a radio signal to reach its destination in a single hop
C. The maximum frequency that allows a radio signal to be absorbed in the lowest ionospheric layer
D. The minimum frequency that allows a radio signal to be absorbed in the lowest ionospheric layer

The answer is A. See answer 3AC-3.1.

3AC-4.1 What is usually the condition of the ionosphere just before sunrise?
A. Atmospheric attenuation is at a maximum
B. Ionization is at a maximum
C. The E layer is above the F layer
D. Ionization is at a minimum

The answer is D. Ionization is at a minimum just before sunrise in the F layer of the atmosphere. In the E layer, it is at a minimum at night.

3AC-4.2 At what time of day does maximum ionization of the ionosphere occur?
A. Dusk   B. Midnight   C. Dawn   D. Midday

The answer is D. This is true because ionization depends upon the sun's radiation, and this is strongest at midday.

3AC-4.3 Which two daytime ionospheric layers combine into one layer at night?
A. E and F1   B. D and E   C. E1 and E2   D. F1 and F2

The answer is D. See answer 3AC-1.1.

3AC-4.4 Minimum ionization of the ionosphere occurs daily at what time?
A. Shortly before dawn   B. Just after noon
C. Just after dusk   D. Shortly before midnight

The answer is A. See answer 3AC-4.1.

3AC-6.1 When two stations are within each other's skip zone on the frequency being used, what mode of propagation would it be desirable to use?
A. Ground wave propagation   B. Sky wave propagation
C. Scatter-mode propagation   D. Ionospheric ducting propagation

The answer is C. The "skip-zone" is the area where there is ordinarily no reception. The sky wave "skips" over it. See Figure 3AC-1.14.

A scatter mode of propagation is used to "penetrate" the skip zone. Scatter refers to the random refraction or reflection of radio waves by irregularities in the earth's atmosphere or in the earth's surface. Scatter signals are invariably weak. However, scatter can provide reliable communications on some VHF frequencies up to several hundred miles if high power transmitters, efficient antennas and sensitive receivers are used.

In the HF bands below 10 meters, a one-hop sky wave returning from the ionosphere, strikes the earth's surface and, depending on the type of surface, may be scattered in a direction back toward the transmitter. It thus fills in the skip zone with a weak signal. This type of scatter is called BACKSCATTER. It is not a dependable propagation mode, but occurs often enough to be used.

3AC-6.3 When is E layer ionization at a maximum?
A. Dawn   B. Midday   C. Dusk   D. Midnight
The answer is B. E layer ionization is maximum at noontime. It disappears in the evening.

3AC-8.1 What is the transmission path of a wave that travels directly from the transmitting antenna to the receiving antenna called?
A. Line of sight   B. The sky wave
C. The linear wave   D. The plane wave
The answer is A. The "line of sight" wave is also referred to as the

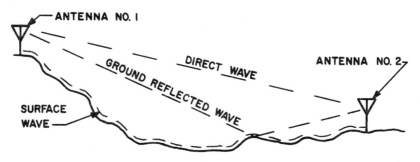

Figure 3AC-8.1. Components of the ground wave.

direct wave. It is a part of the **ground wave.** The other parts of the ground wave are the **surface wave** and the **ground reflected wave.** All three parts of the ground wave are shown in figure 3AC-8.1.

3AC-8.2 How are VHF signals within the range of the visible horizon propagated?
A. By sky wave   B. By direct wave
C. By plane wave   D. By geometric wave
The answer is B. See answer 3AC-8.1.

3AC-9.1 Ducting occurs in which region of the atmosphere?
A. F2   B. Ionosphere   C. Stratosphere   D. Troposphere
The answer is D. The troposphere is that part of the earth's atmosphere that extends from the earth's surface to a height of 5 to 8 miles.

3AC-9.2 What effect does tropospheric bending have on 2 meter radio waves?
A. It increases the distance over which they can be transmitted
B. It decreases the distance over which they can be transmitted
C. It tends to garble 2-meter phone transmissions
D. It reverses the sideband of 2-meter phone transmissions

The answer is A. The air temperature and density decrease upward from the earth's surface into the troposphere. A signal entering the troposphere will bend because of the air changes. This bending is called "tropospheric bending" and it causes a signal to be received at a point considerably beyond the "line of sight". Tropospheric bending is most pronounced in the VHF region, especially 144 to 148 MHz.

3AC-9.3 What atmospheric phenomenon causes tropospheric ducting of radio waves?
A. A very low pressure area
B. An aurora to the north
C. Lightning between the transmitting and receiving station
D. A temperature inversion

The answer is D. Normally, the air temperature decreases upward from the earth's surface. However, under certain conditions, the warmer air may occur at an altitude between layers of cold air. This produces an enclosed air "duct", similar to a wave guide. A signal, "trapped" in this type of air duct, can travel hundreds and thousands of miles. Tropospheric ducting occurs more on the 2 meter band than on the other amateur bands.

3AC-9.4 Tropospheric ducting occurs as a result of what phenomenon?
A. A temperature inversion   B. Sun spots
C. An aurora to the north
D. Lightning between the transmitting and receiving station

The answer is A. The phenomenon of warmer air layers occurring at a higher altitude than cooler air is also referred to as "temperature inversions" or "inversion layers". See answer 3AC-9.3.

3AC-9.5 What atmospheric phenomenon causes VHF radio waves to be propagated several hundred miles through stable air masses over oceans?
A. Presence of a maritime polar air mass
B. A widespread temperature inversion
C. An overcast of cirriform clouds
D. Atmospheric pressure of roughly 29 inches of mercury or higher

The answer is B. See answers 3AC-9.2 through 3AC-9.4.

3AC-9.6 In what frequency range does tropospheric ducting occur most often?
A. LF   B. MF   C. HF   D. VHF

The answer is D. It occurs in the VHF range, especially 144 MHz and up.

SUBELEMENT 3AD
AMATEUR RADIO PRACTICE
(4 questions)

3AD-1.1 Where should the green wire in an AC line cord be attached in a power supply?
A. To the fuse
B. To the "hot" side of the power switch
C. To the chassis
D. To the meter
The answer is C. The green wire is the "equipment ground" wire and should be connected to the chassis of the equipment.

3AD-1.2 Where should the black (or red) wire in a three-wire line cord be attached in a power supply?
A. To the filter capacitor   B. To the DC ground
C. To the chassis   D. To the fuse
The answer is D. The black, or red or blue wire is the "hot" lead, and the "hot" lead must go to the fuse.

3AD-1.3 Where should the white wire in a three-wire line cord be attached in a power supply?
A. To the fuse
B. To one side of the transformer's primary winding
C. To the black wire
D. To the rectifier junction
The answer is B. The white wire is the "system" neutral conductor. The white wire and the "hot" wire, deliver the 117 volts in a 117 volt system.

3AD-1.4 Why is the retaining screw in one terminal of a light socket made of brass, while the other one is silver colored?
A. To prevent galvanic action   B. To indicate correct polarity
C. To better conduct current   D. To reduce skin effect
The answer is B. The brass colored screw is the "hot" terminal and the silver screw is the "neutral" terminal.

3AD-2.1 How much electrical current flowing through the human body is usually fatal?
A. As little as 100 milliamperes may be fatal
B. Approximately 10 amperes is required to be fatal
C. More than 20 amperes is needed to kill a human being
D. No amount of current will harm you. Voltages of over 2000 volts are always fatal, however.
The answer is A. It should be pointed out that currents as low as 10 to 15 ma. will cause severe shock and muscle paralysis.

3AD-2.2 What is the minimum voltage considered to be dangerous to humans?
A. 30 volts   B. 100 volts   C. 1000 volts   D. 2000 volts
The answer is A. Contrary to common belief, a low voltage can cause a fatal shock if good contact is made with the source and there is adequate

current available.

3AD-2.3 Where should the main power-line switch for a high voltage power supply be situated?
A. Inside the cabinet, to interrupt power when the cabinet is opened
B. On the rear panel of the high-voltage supply
C. Where it can be seen and reached easily
D. This supply should not be switch-operated

The answer is C. The main power switch should always be visible and accessible in the event of a safety problem. Also, it should be turned off when the station is not in use.

3AD-2.5 How much electrical current flowing through the human body is usually painful?
A. As little as 50 milliamperes may be painful
B. Approximately 10 amperes is required to be painful
C. More than 20 amperes is needed to be painful to a human being
D. No amount of current will be painful. Voltages of over 2000 volts are always painful, however

The answer is A. Actually, we can have pain with as little as 5 to 10 ma.

3AD-5.2 Where in the antenna transmission line should a peak-reading wattmeter be attached to determine the transmitter output power?
A. At the transmitter output   B. At the antenna feedpoint
C. One-half wavelength from the antenna feedpoint
D. One-quarter wavelength from the transmitter output

The answer is A. If we put the meter at the antenna feedpoint, we would be reading the transmitter output power, LESS the transmission line loss.

3AD-5.3 If a directional rf wattmeter indicates 90 watts forward power and 10 watts reflected power, what is the actual transmitter output power?
A. 10 watts   B. 80 watts   C. 90 watts   D. 100 watts

The answer is B. We subtract the reflected power from the forward power to arrive at the actual transmitter output power.
90 watts - 10 watts = 80 watts.

3AD-5.4 If a directional rf wattmeter indicates 96 watts forward power and 4 watts reflected power, what is the actual transmitter output power?
A. 80 watts   B. 88 watts   C. 92 watts   D. 100 watts

The answer is C. We subtract the reflected power from the forward power to arrive at the actual transmitter output power.
96 watts - 4 watts = 92 watts.

3AD-7.1 What is a multimeter?
A. An instrument capable of reading SWR and power
B. An instrument capable of reading resistance, capacitance, and inductance
C. An instrument capable of reading resistance and reactance
D. An instrument capable of reading voltage, current, and resistance

The answer is D. A multimeter is an ammeter, voltmeter and ohmmeter combined into a single case, using one meter movement. A simple multimeter will suffice for most electronic servicing encountered outside of the laboratory.

3AD-7.2 How can the range of a voltmeter be extended?
A. By adding resistance in series with the circuit under test
B. By adding resistance in parallel with the circuit under test
C. By adding resistance in series with the meter
D. By adding resistance in parallel with the meter
The answer is C. The range of a voltmeter can be extended by adding resistors in series with the voltmeter. These are called "multipliers".

3AD-7.3 How is a voltmeter typically connected to a circuit under test?
A. In series with the circuit   B. In parallel with the circuit
C. In quadrature with the circuit   D. In phase with the circuit
The answer is B. A voltmeter is always connected in parallel with the circuit. The higher the internal resistance of the voltmeter, the more accurate is the measurement.

3AD-7.4 How can the range of an ammeter be extended?
A. By adding resistance in series with the circuit under test
B. By adding resistance in parallel with the circuit under test
C. By adding resistance in series with the meter
D. By adding resistance in parallel with the meter
The answer is D. The range of an ammeter can be extended by adding shunts in parallel with the ammeter. A shunt is a low value resistance. An ammeter is connected in series with the circuit that it is measuring.

3AD-8.1 What is a marker generator?
A. A high-stability oscillator that generates a series of reference signals at known frequency intervals
B. A low-stability oscillator that "sweeps" through a band of frequencies
C. An oscillator often used in aircraft to determine the craft's location relative to the inner and outer markers at airports
D. A high-stability oscillator whose output frequency and amplitude can be varied over a wide range
The answer is A. A marker generator is a simple RF generator that can be adjusted to put out one or more specific frequency signals. It can be used in conjunction with a sweep generator to indicate frequency marks on the response curve shown on an oscilloscope. The marker generator can be part of the sweep generator, or it can be a separate generator hooked up in parallel with the sweep generator. See answer 3AD-8.4.

3AD-8.2 What piece of test equipment provides a variable- frequency signal which can be used to check the frequency response of a circuit?
A. Frequency counter   B. Distortion analyzer
C. Deviation meter   D. Signal generator
The answer is D. See answer to question 3AD-8.1.

3AD-8.3 What type of circuit is used to inject a frequency calibration signal into a communications receiver?
A. A product detector
B. A receiver incremental tuning circuit
C. A balanced modulator   D. A crystal calibrator

The answer is D. A crystal calibrator is a crystal-controlled oscillator that generates accurate, stable signals at a given frequency and its harmonic frequencies. It is used to insure the dial accuracy of the receiver.

3AD-8.4 How is a marker generator used?
A. To calibrate the tuning dial on a receiver
B. To calibrate the volume control on a receiver
C. To test the amplitude linearity of an SSB transmitter
D. To test the frequency deviation of an FM transmitter

The answer is A. A marker generator is an RF generator that generates an RF signal. Its purpose is to pinpoint specific frequencies on the response curve. The output frequency of the marker generator can be varied by a knob on the panel. One type of marker generator can generate a number of harmonically related signals at the same time.

3AD-8.5 When adjusting a transmitter filter circuit, what device is connected to the transmitter output?
A. Multimeter   B. Litz wires   C. Receiver   D. Dummy antenna

The answer is D. We use a dummy antenna instead of the actual antenna because we do not want to radiate RF signals, with its consequent interference, while we are adjusting the transmitter.

3AD-11.1 What is a reflectometer?
A. An instrument used to measure signals reflected from the ionosphere
B. An instrument used to measure standing wave ratio
C. An instrument used to measure transmission-line impedance
D. An instrument used to measure radiation resistance

The answer is B. A reflectometer can be used to measure the SWR (standing wave ratio) in an antenna feedline. It can also measure the relative power output of the transmitter and can, therefore, be used in tuning the transmitter.

3AD-11.2 For best accuracy when adjusting the impedance match between an antenna and feedline, where should the match-indicating device be inserted?
A. At the antenna feedpoint   B. At the transmitter
C. At the midpoint of the feedline   D. Anywhere along the feedline

The answer is A. It should be inserted at the antenna end of the feedline. Putting it any place else gives a less accurate, although sometimes acceptable, reading.

3AD-11.3 What is the device that can indicate an impedance mismatch in an antenna system?
A. A field strength meter   B. A set of lecher wires
C. A wavemeter   D. A reflectometer

The answer is D. See answer 3AD-11.1.

3AD-11.4 What is a reflectometer?
A. An instrument used to measure signals reflected from the ionosphere
B. An instrument used to measure standing wave ratio
C. An instrument used to measure transmission-line impedance
D. An instrument used to measure radiation resistance

The answer is B. See answer 3AD-11.1.

3AD-11.5 Where should a reflectometer be inserted into a long antenna transmission line in order to obtain the most valid standing wave ratio indication?
A. At any quarter wavelength interval along the transmission line
B. At the receiver end
C. At the antenna end
D. At any even half-wavelength interval along the transmission line

The answer is C. It should be inserted at the antenna end of the line; that is, where the transmission line connects to the antenna.

3AD-12.1 What result might be expected when using a speech processor with an emission J3E transmitter?
A. A lower plate-current reading
B. A less natural-sounding voice
C. A cooler operating power supply
D. Greater PEP output

The answer is B. It would be less natural sounding, compared to AM.

3AD-14.1 What is a transmatch?
A. A device for varying the resonant frequency of an antenna
B. A device for varying the impedance presented to the transmitter
C. A device for varying the tuning rate of the transmitter
D. A device for varying the electrical length of an antenna

The answer is B. A transmatch is an antenna tuning unit that is inserted between the transmitter and the transmission line leading to the antenna. It consists of a network of coil-capacitor circuits, and its primary function is to match the impedance of the transmitter to the impedance of its load (the transmission line and antenna).

3AD-14.2 What is a balanced line?
A. Feed line with one conductor connected to ground
B. Feed line with both conductors connected to ground to balance out harmonics
C. Feed line with the outer conductor connected to ground at even intervals
D. Feed line with neither conductor connected to ground

The answer is D. There are two basic types of transmission lines. One is called parallel or balanced. The other is unbalanced or coaxial cable. Balanced lines are further subdivided into two different types. One is the open wire type using insulating spacers; the other is the common television

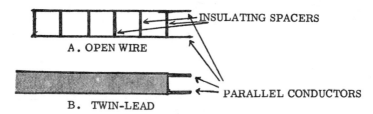

Figure 3AD-14.2. Parallel conductor transmission lines.

twin-lead type that uses continuous insulation. See Figure 3AD-14.2 (A and B). Both types of balanced lines consist of two similar conductors parallel to each other and separated by some form of insulation.

3AD-14.3 What is an unbalanced line?
A. Feed line with neither conductor connected to ground
B. Feed line with both conductors connected to ground to suppress harmonics
C. Feed line with one conductor connected to ground
D. Feed line with the outer conductor connected to ground at uneven intervals

The answer is C. **Coaxial cable** is an example of an unbalanced line. It consists of an inner conductor surrounded by a round flexible insulator. Around this insulator, there is a concentric metallic conductor made of flexible wire braid. A weatherproof vinyl sheath surrounds the braid. The outer conductor, which is connected to ground, acts as a shield, preventing spurious and harmonic radiation from the transmission line. Coaxial cable is illustrated in Figure 3AD-14.3.

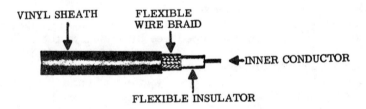

Figure 3AD-14.3. Coaxial Cable.

3AD-14.4 What is a balun?
A. A device for using an unbalanced line to supply power to a balanced load, or vice versa
B. A device to match impedances between two coaxial lines
C. A device used to connect a microphone to a balanced modulator
D. A counterbalance used with an azimuth/elevation rotator system

The answer is A. A balun is a type of transformer that is used to transfer power between a balanced device and an unbalanced device. A typical use of a balun is at the antenna input, where an unbalanced transmission line feeds a balanced dipole antenna. Without the balun, antenna currents would be induced in the shield and cause the transmission line to radiate. This would tend to destroy the radiation pattern of the antenna. The balun remedies this situation by isolating the balanced antenna from the unbalanced line. At the same time, it provides the maximum transfer of RF power between the transmission line and the antenna.

Baluns are also used for impedance matching. They can effect optimum energy transfer from a source having a certain impedance to a load having a different impedance.

3AD-14.5 What is the purpose of an antenna matching circuit?
A. To measure the impedance of the antenna
B. To compare the radiation patterns of two antennas
C. To measure the SWR of an antenna

D. To match impedances within the antenna system
The answer is D. Impedances must be matched in order to transfer maximum power between the transmitter and the antenna system.

3AD-14.8 How is a transmatch used?
A. It is connected between a transmitter and an antenna system, and tuned for minimum SWR at the transmitter
B. It is connected between a transmitter and an antenna system and tuned for minimum SWR at the antenna
C. It is connected between a transmitter and an antenna system, and tuned for minimum impedance
D. It is connected between a transmitter and a dummy load, and tuned for maximum output power

The answer is A. A transmatch is an antenna tuning unit that is inserted between the transmitter and the transmission line leading to the antenna. It consists of a network of coil-capacitor circuits, and its primary function is to match the impedance of the transmitter to the impedance of its load (the transmission line and antenna).

The transmatch should be adjusted for minimum reading in the reflected direction, and for maximum reading in the forward direction.

3AD-16.1 What is a dummy antenna?
A. An isotropic radiator
B. A nonradiating load for a transmitter
C. An antenna used as a reference for gain measurements
D. The image of an antenna, located below ground

The answer is B. A dummy antenna is a pure resistance load that presents the same resistance and power dissipation to the final stage as the antenna does. A dummy antenna is located right in the amateur station's operating room, and converts the transmitter's output energy into heat. This prevents the signals from getting out on the air during the testing and tuning procedures. After these procedures have been completed, using the dummy load, the dummy load is switched out and the antenna is switched in. The final "touch-up" tuning, which should take very little time, can then be done.

3AD-16.2 Of what materials may a dummy antenna be made?
A. A wire-wound resistor
B. A noninductive resistor
C. A diode and resistor combination
D. A coil and capacitor combination

The answer is B. The dummy load should be a pure resistance. It may be difficult to obtain high power resistors that are non-inductive. It may therefore be necessary to put together many low power resistors in a series-parallel configuration to yield the correct resistance and power rating.

3AD-16.3 What station accessory is used in place of an antenna during transmitter tests so that no signal is radiated?
A. A Transmatch   B. A dummy antenna
C. A low-pass filter   D. A decoupling resistor

The answer is B. See answers 3AD-16.1 and 3AD-16.2.

3AD-16.4 What is the purpose of a dummy load?
A. To allow off-the-air transmitter testing
B. To reduce output power for QRP operation
C. To give comparative signal reports
D. To allow Transmatch tuning without causing interference
The answer is A. See answers 3AD-16.1 and 3AD-16.2.

3AD-16.5 How many watts should a dummy load for use with a 100 watt emission J3E transmitter with 50 ohm output be able to dissipate?
A. A minimum of 100 watts continuous
B. A minimum of 141 watts continuous
C. A minimum of 175 watts continuous
D. A minimum of 200 watts continuous
The answer is A. The dummy load must be able to dissipate the full power output of the transmitter because it may receive the full power output of the transmitter during testing and tuning.

3AD-17.1 What is an S-meter?
A. A meter used to measure sideband suppression
B. A meter used to measure spurious emissions from a transmitter
C. A meter used to measure relative signal strength in a receiver
D. A meter used to measure solar flux
The answer is C. Many communications receivers have an "S" meter built into them. The S meter indicates the relative signal strength of the received signal.

There are many types of S meter circuits. The simplest type consists of a rectifier and DC microammeter that reads the signal at the output of the last IF amplifier. In addition to being a signal strength indicator, the S meter can also be used as an output indicator in tuning up the receiver.

It is important to keep in mind the fact that the S meter is not an accurate signal strength meter, but rather a means of obtaining relative signal strength indications.

3AD-18.1 For the most accurate readings of transmitter output power, where should the rf wattmeter be inserted?
A. The wattmeter should be inserted and the output measured one-quarter wavelength from the antenna feedpoint
B. The wattmeter should be inserted and the output measured one-half wavelength from the antenna feedpoint
C. The wattmeter should be inserted and the output power measured at the transmitter antenna jack
D. The wattmeter should be inserted and the output power measured at the Transmatch output
The answer is C. This tells us how much power is coming out of the transmitter. It does not tell us how much power is at the antenna.

3AD-18.2 At what line impedance are rf wattmeters usually designed to operate?
A. 25 ohms   B. 50 ohms   C. 100 ohms   D. 300 ohms
The answer is B. For the most accurate readings, the impedance of the wattmeter should be equal to the transmitter output impedance. Since the transmitter output impedance is usually 50 ohms, wattmeters are generally

designed to operate at 50 ohms.

3AD-18.3 What is a directional wattmeter?
A. An instrument that measures forward or reflected power
B. An instrument that measures the directional pattern of an antenna
C. An instrument that measures the energy consumed by the transmitter
D. An instrument that measures thermal heating in a load resistor

The answer is A. A switch on the directional wattmeter can make it read forward or reflected power. In order to know the actual power that is delivered to the load, the reflected power must be subtracted from the forward power.

SUBELEMENT 3AE
ELECTRICAL PRINCIPLES
(2 questions)

3AE-2.1 What is meant by the term resistance?
A. The opposition to the flow of current in an electric circuit containing inductance
B. The opposition to the flow of current in an electric circuit containing capacitance
C. The opposition to the flow of current in an electric circuit containing reactance
D. The opposition to the flow of current in an electric circuit that does not contain reactance

The answer is D. Resistance is the opposition that a material offers to current. Resistance is different from reactance. Reactance is the opposition that an inductor and capacitor offer to alternating current.

3AE-2.2 What is the primary function of a resistor?
A. To store an electric charge
B. To store a magnetic field
C. To match a high-impedance source to a low-impedance load
D. To limit the current in an electric circuit

The answer is D. A resistor can be used for many purposes. In addition to limiting current it can be used in conjunction with other resistors to divide voltages.

3AE-2.3 What is a variable resistor?
A. A resistor with a slide or contact that makes the resistance adjustable
B. A device that can transform a variable voltage into a constant voltage
C. A resistor that changes value when an ac voltage is applied to it
D. A resistor that changes value when it is heated

The answer is A. As the contact moves along the resistance material, the amount of resistance that is available to the circuit changes.

3AE-2.4 Why do resistors generate heat?
A. They convert electrical energy to heat energy
B. They exhibit reactance
C. Because of skin effect
D. To produce thermionic emission

The answer is A. When electrons flow through the resistor, there is friction between the moving electrons and the molecules of the resistor. This friction causes heat to be generated. The higher the current, the greater is the heat.

3AE-4.1 What is an inductor?
A. An electronic component that stores energy in an electric field
B. An electronic component that converts a high voltage to a lower voltage
C. An electronic component that opposes dc while allowing ac to pass
D. An electronic component that stores energy in a magnetic field

The answer is D. See answer 3AE-4.3.

3AE-4.2 What factors determine the amount of inductance in a coil?
A. The type of material used in the core, the diameter of the core and whether the coil is mounted horizontally or vertically
B. The diameter of the core, the number of turns of wire used to wind the coil and the type of metal used in the wire
C. The type of material used in the core, the number of turns used to wind the core and the frequency of the current through the coil
D. The type of material used in the core, the diameter of the core, the length of the coil and the number of turns of wire used to wind the coil

The answer is D. The inductance of a coil varies directly with the square of the number of turns of wire in the coil, the permeability of the core material, and the cross sectional area of the core. It varies inversely with the length of the coil.

3AE-4.3 What are the electrical properties of an inductor?
A. An inductor stores a charge electrostatically and opposes a change is voltage
B. An inductor stores a charge electrochemically and opposes a change in current
C. An inductor stores a charge electromagnetically and opposes a change in current
D. An inductor stores a charge electromechanically and opposes a change in voltage

The answer is C. An inductor is a circuit component that has the property of inducing a voltage in itself when the current through it varies. This induced voltage opposes the variations or changes in the current, and is called a counter e.m.f.

When a current flows through an inductor, a magnetic field is built up around the inductor. The inductor actually "stores" energy in the magnetic field.

3AE-4.4 What is an inductor core?
A. The central portion of a coil; may be made from air, iron, brass or other material
B. A tight coil of wire used in a transformer
C. An insulating material placed between the plates of an inductor
D. The point at which an inductor is tapped to produce resonance

The answer is A. An iron core coil has higher inductance than an air core coil. This is because iron has a greater permeability than air and allows for a larger magnetic flux.

3AE-4.5 What are the component parts of a coil?
A. The wire in the winding and the core material
B. Two conductive plates and an insulating material
C. Two or more layers of silicon material
D. A donut-shaped iron core and a layer of insulating tape

The answer is A. See answers 3AE-4.1 through 3AE-4.4.

3AE-5.1 What is a capacitor?
A. An electronic component that stores energy in a magnetic field

B. An electronic component that stores energy in an electric field
C. An electronic component that converts a high voltage to a lower voltage
D. An electronic component that converts power into heat

The answer is B. A capacitor consists of two conductors, separated by an insulator (dielectric). The conductors can be aluminum plates, copper plates, tin foil, or other similar materials. The insulators can be air, mica, waxed paper, or similar materials.

3AE-5.2 What factors determine the amount of capacitance in a capacitor?
A. The dielectric constant of the material between the plates, the area of one side of one plate, the separation between the plates and the number of plates
B. The dielectric constant of the material between the plates, the number of plates and the diameter of the leads connected to the plates
C. The number of plates, the spacing between the plates and whether the dielectric material is N type or P type
D. The dielectric constant of the material between the plates, the surface area of one side of one plate, the number of plates and the type of material used for the protective coating

The answer is A. The capacitance is directly proportional to the area and number of the plates and the dielectric constant of the material between the plates. It is inversely proportional to the separation between the plates.

3AE-5.3 What are the electrical properties of a capacitor?
A. A capacitor stores a charge electrochemically and opposes a change in current
B. A capacitor stores a charge electromagnetically and opposes a change in current
C. A capacitor stores a charge electromechanically and opposes a change in voltage
D. A capacitor stores a charge electrostatically and opposes a change in voltage

The answer is D. A capacitor builds up a counter EMF which opposes alternating current. This opposition is referred to as capacitive reactance and is measured in ohms.

3AE-5.4 What is a capacitor dielectric?
A. The insulating material used for the plates
B. The conducting material used between the plates
C. The ferrite material that the plates are mounted on
D. The insulating material between the plates

The answer is D. Some common capacitor dielectrics in use are mica, ceramic, paper, air and glass.

3AE-5.5 What are the component parts of a capacitor?
A. Two or more conductive plates with an insulating material between them
B. The wire used in the winding and the core material
C. Two or more layers of silicon material

ELECTRICAL PRINCIPLES TE-4

D. Two insulating plates with a conductive material between them
The answer is A. See answers 3AE-5.1 through 3AE-5.4.

3AE-7.1 What is an ohm?
A. The basic unit of resistance
B. The basic unit of capacitance
C. The basic unit of inductance
D. The basic unit of admittance

The answer is A. The basic unit of resistance is the OHM. The ohm is defined as the amount of resistance inherent in 1,000 feet of #10 copper wire. For example, 3,000 feet of #10 copper wire has 3 ohms of resistance.

3AE-7.3 What is the unit measurement of resistance?
A. Volt   B. Ampere   C. Joule   D. Ohm

The answer is D. See answers 3AE-7.1 and 3BE-7.2.

3AE-8.1 What is a microfarad?
A. A basic unit of capacitance equal to 10^{-6} farads
B. A basic unit of capacitance equal to 10^{-12} farads
C. A basic unit of capacitance equal to 10^{-2} farads
D. A basic unit of capacitance equal to 10^{6} farads

The answer is A. A farad, abbreviated "f", is the basic unit of capacitance. A microfarad, abbreviated "mf", is one millionth of a farad. A picofarad, abbreviated "pf", is a millionth of a millionth of a farad. These relationships can be stated in decimal form as follows:

1 f = 1,000,000 mf.   1 mf = .000001 f.
1 f = 1,000,000,000,000 pf.   1 pf = .000000000001 f.
1 mf = 1,000,000 pf.   1 pf = .000001 mf.

3AE-8.2 What is a picofarad?
A. A basic unit of capacitance equal to 10^{-6} farads
B. A basic unit of capacitance equal to 10^{-12} farads
C. A basic unit of capacitance equal to 10^{-2} farads
D. A basic unit of capacitance equal to 10^{6} farads

The answer is B. See answer 3AE-8.1.

3AE-8.3 What is a farad?
A. The basic unit of resistance   B. The basic unit of capacitance
C. The basic unit of inductance   D. The basic unit of admittance

The answer is B. See answer 3AE-8.1.

3AE-8.4 What is the basic unit of capacitance?
A. Farad   B. Ohm   C. Volt   D. Ampere

The answer is A. See answer 3AE-8.1.

3AE-9.1 What is a microhenry?
A. A basic unit of inductance equal to 10^{-12} henrys
B. A basic unit of inductance equal to 10^{-3} henrys
C. A basic unit of inductance equal to 10^{6} henrys
D. A basic unit of inductance equal to 10^{-6} henrys

The answer is D. The "Henry" is the basic unit of inductance. A millihenry is equal to one thousandth of a henry. A microhenry is equal to one millionth of a henry.

3AE-9.2 What is a millihenry?
A. A basic unit of inductance equal to 10^{-6} henrys
B. A basic unit of inductance equal to 10^{-12} henrys
C. A basic unit of inductance equal to 10^{-3} henrys
D. A basic unit of inductance equal to 10^{6} henrys
   The answer is C. See answer 3AE-9.1.

3AE-9.3 What is a henry?
A. The basic unit of resistance   B. The basic unit of capacitance
C. The basic unit of inductance   D. The basic unit of admittance
   The answer is C. See answer 3AE-9.1.

3AE-9.4 What is the basic unit of inductance?
A. Coulomb   B. Farad   C. Ohm   D. Henry
   The answer is D. See answer 3AE-9.1.

3AE-11.1 How is the current in a dc circuit calculated when the voltage and resistance are known?
A. I = E/R   B. P = I x E   C. I = R x E   D. I = E x R
   The answer is A. This is the basic formula for Ohm's Law. It tells us that the current is equal to the voltage divided by the resistance.

3AE-11.2 What is the input resistance of a load when a 12-volt battery supplies 0.25-amperes to it?
A. 0.02 ohms   B. 3 ohms   C. 48 ohms   D. 480 ohms
   The answer is C. In order to find the resistance of a circuit when the voltage and current are known, we use the basic Ohm's Law formula, but we transpose the current and the resistance.

$$\text{Therefore, } R = \frac{E}{I}$$

We then substitute the voltage and current to find the resistance.

$$R = \frac{12}{0.25} = 48 \text{ ohms}$$

Another form of Ohm's Law is E = I x R. We use this formula when we know the current and resistance, and wish to find the voltage.

3AE-11.3 The product of the current and what force gives the electrical power in a circuit?
A. Magnetomotive force   B. Centripetal force
C. Electrochemical force   D. Electromotive force
   The answer is D. The product of the current and the electromotive force, or voltage, gives us the electrical power in a circuit. We can state this mathematically, as follows:

$$P = I \times E \quad \text{where:} \quad \text{P is the power in watts,}$$
$$\text{I is the current in amperes and}$$
$$\text{E is the voltage in volts}$$

ELECTRICAL PRINCIPLES   TE-6

3AE-11.4 What is Ohm's Law?
A. A mathematical relationship between resistance, current and applied voltage in a circuit
B. A mathematical relationship between current, resistance and power in a circuit
C. A mathematical relationship between current, voltage, and power in a circuit
D. A mathematical relationship between resistance, voltage and power in a circuit

The answer is A. Ohm's Law states that the current is directly proportional to the voltage and is inversely proportional to the resistance. The three forms of Ohm's Law are:

$$A)\ I = \frac{E}{R} \qquad B)\ E = I \times R \qquad C)\ R = \frac{E}{I}$$

where: I is the current in amperes
E is the voltage in volts
R is the resistance in ohms

3AE-11.5 What is the input resistance of a load when a 12-volt battery supplies 0.15-amperes to it?
A. 8 ohms   B. 80 ohms   C. 100 ohms   D. 800 ohms

The answer is B. We use formula C of 3AE-11.4 to solve this problem.

$$R = \frac{E}{I} = \frac{12}{0.15} = 80\ ohms$$

3AE-12.2 In a series circuit composed of a voltage source and several resistors, what determines the voltage drop across any particular resistor?
A. It is equal to the source voltage
B. It is equal to the source voltage, divided by the number of series resistors in the circuit
C. The larger the resistor's value, the greater the voltage drop across that resistor
D. The smaller the resistor's value, the greater the voltage drop across that resistor

The answer is C. The voltage drop across any resistor is equal to the current through the resistor, multiplied by the resistance value of the resistor. When resistors are in series across a source, the voltage drop across any particular resistor is directly proportional to the value of the resistance. The numerical value of the voltage drop across a particular resistor is equal to that resistor's resistance, divided by the total resistance, multiplied by the source voltage.

3AE-13.4 How is power calculated when the current and voltage in a circuit are known?
A. E = I x R   B. P = I x E   C. P = I x I / R   D. P = E / I

The answer is B. See answer 3AE-11.3.

3AE-14.8 When 120-volts is measured across a 4700 ohm resistor, approximately how much current is flowing through it?
A. 39 amperes   B. 3.9 ampere   C. 0.26 ampere   D. 0.026 ampere

The answer is D. We use Ohm's Law to solve this problem.

$$I = \frac{E}{R} = \frac{120}{4700} = 0.026 \text{ A}$$

3AE-14.9 When 120-volts is measured across a 47000 ohm resistor, approximately how much current is flowing through it?
A. 392 A   B. 39.2 A   C. 26 mA   D. 2.6 mA
The answer is D. We use Ohm's Law to solve this problem.

$$I = \frac{E}{R} = \frac{120}{47000} = 0.0026 \text{ A or } 2.6 \text{ mA}$$

3AE-14.10 When 12-volts is measured across a 4700 ohm resistor, approximately how much current is flowing through it?
A. 2.6 mA   B. 26 mA   C. 39.2 A   D. 392 A
The answer is A. We use Ohm's Law to solve this problem.

$$I = \frac{E}{R} = \frac{12}{4700} = 0.0026 \text{ A or } 2.6 \text{ mA}$$

3AE-14.11 When 12-volts is measured across a 47000 ohm resistor, approximately how much current is flowing through it?
A. 255 µA   B. 255 mA   C. 3917 mA   D. 3917 A
The answer is A. We use Ohm's Law to solve this problem.

$$I = \frac{E}{R} = \frac{12}{47000} = 0.000255 \text{ A} = 255 \text{ microamperes (µA)}$$

SUBELEMENT 3AF
CIRCUIT COMPONENTS
(2 questions)

3AF-1.1 How can a carbon resistor's electrical tolerance rating be found?
A. By using a wavemeter
B. By using the resistor's color code
C. By using Thevenin's theorem for resistors
D. By using the Baudot code
The answer is B. A carbon resistor generally contains four bands. The first three bands starting at the end, give the value of the resistance. The fourth band gives the tolerance. If the fourth band is gold, the tolerance is 5%. If it is silver, the tolerance is 10%. If there is no fourth band, the tolerance is 20%.

3AF-1.2 Why would a large size resistor be substituted for a smaller one of the same resistance?
A. To obtain better response
B. To obtain a higher current gain
C. To increase power dissipation capability
D. To produce a greater parallel impedance
The answer is C. A physically large size resistor would be substituted for a small size resistor of the same resistance value if the normal power dissipated in the original resistance was greater than it was able to handle.

3AF-1.3 What do the first three color bands on a resistor indicate?
A. The value of the resistor in ohms
B. The resistance tolerance in percent
C. The power rating in watts
D. The value of the resistor in henrys
The answer is A. The first band from the end represents the first figure of the resistance value. The second band represents the second figure of the resistance value. The third band represents the number of zeros after the first two numbers. For instance, color bands red-green-orange indicate a value of 25,000 ohms.

3AF-1.4 What does the fourth color band on a resistor indicate?
A. The value of the resistor in ohms
B. The resistance tolerance in percent
C. The power rating in watts
D. The resistor composition
The answer is B. See answers 3AF-1.1 and 3AF-1.6.

3AF-1.6 When the color bands on a group of resistors indicate that they all have the same resistance, what further information about each resistor is needed in order to select those that have nearly equal value?
A. The working voltage rating of each resistor
B. The composition of each resistor

C. The tolerance of each resistor
D. The current rating of each resistor

The answer is C. In order to insure that the resistors have nearly equal value, we must look to the fourth color which tells us the tolerance. For instance, if a thousand ohm resistor has a gold band indicating 5% tolerance, its value lies anywhere between 950 and 1050 ohms. If another thousand ohm resistor has a silver band, its resistance varies anywhere from 900 ohms to 1100 ohms. If still another thousand ohm resistor has no band, its value can be anywhere between 800 and 1200 ohms. Therefore, if we wish to choose resistors of equal value, we must insist upon a fourth gold color band, which would indicate that they are all close in value to each other. Resistors are also made of closer tolerances, within 1 or 2%. These resistors are, of course, more expensive than the others.

3AF-2.1 As the plate area of a capacitor is increased, what happens to its capacitance?
A. Decreases
B. Increases
C. Stays the same
D. Becomes voltage dependent

The answer is B. As the plate area of a capacitor increases, the capacity of the capacitor also increases. As the spacing between the plates of a capacitor increases, the capacitance value decreases.

3AF-2.2 As the plate spacing of a capacitor is increased, what happens to its capacitance?
A. Increases
B. Stays the same
C. Becomes voltage dependent
D. Decreases

The answer is D. See answer 3AF-2.1.

3AF-2.3 What is an electrolytic capacitor?
A. A capacitor whose plates are formed on a thin ceramic layer
B. A capacitor whose plates are separated by a thin strip of mica insulation
C. A capacitor whose dielectric is formed on one set of plates through electrochemical action
D. A capacitor whose value varies with applied voltage

The answer is C. A basic electrolytic capacitor consists of two pieces of aluminum, separated by a dielectric. The dielectric is a very thin film of aluminum oxide which is produced by chemical action. All electrolytic capacitors have a positive and a negative terminal, and this polarity must be observed. The electrolytic capacitor has a large amount of capacity in a small physical area because of its very thin dielectric.

3AF-2.4 What is a paper capacitor?
A. A capacitor whose plates are formed on a thin ceramic layer
B. A capacitor whose plates are separated by a thin strip of mica insulation
C. A capacitor whose plates are separated by a layer of paper
D. A capacitor whose dielectric is formed on one set of plates through electrochemical action

The answer is C. We usually describe a capacitor by its dielectric. A paper capacitor is one whose dielectric is waxed paper.

3AF-2.5 What factors must be considered when selecting a capacitor for

a circuit?
A. Type of capacitor, capacitance and voltage rating
B. Type of capacitor, capacitance and the kilowatt-hour rating
C. The amount of capacitance, the temperature coefficient and the KVA rating
D. The type of capacitor, the microscopy coefficient and the temperature coefficient

The answer is A. We must be certain that the voltage rating is not exceeded; otherwise, the capacitor may be destroyed.

3AF-2.8 How are the characteristics of a capacitor usually specified?
A. In volts and amperes
B. In microfarads and volts
C. In ohms and watts
D. In millihenries and amperes

The answer is B. A capacitor is usually specified by its amount of capacitance in farads, microfarads or picofarads, and in the maximum amount of voltage that it can handle without harming itself.

3AF-3.1 What can be done to raise the inductance of 5 microhenry air-core coil to a 5-millihenry coil with the same physical dimensions?
A. The coil can be wound on a non-conducting tube
B. The coil can be wound on an iron core
C. Both ends of the coil can be brought around to form the shape of a doughnut, or toroid.
D. The coil can be made of a heavier-gauge wire

The answer is B. In order to increase the inductance of an air core coil, we can either increase the number of turns or insert, into the coil, a higher permeability core, such as iron. Since the question indicates that we are to have the same physical dimensions, we would have to add an iron, or similar type of core, to increase its inductance without increasing the overall dimensions.

3AF-3.2 Describe an inductor.
A. A semiconductor in a conducting shield
B. Two parallel conducting plates
C. A straight wire conductor mounted inside a Faraday shield
D. A coil of conducting wire

The answer is D. Generally speaking, when we refer to an inductor, we refer to a coil of wire with or without an iron or similar type of core. However, even a piece of wire or a piece of wire with a half turn, has a certain amount of inductance. To get more inductance, we simply increase the number of turns. See answer 3AF-3.1.

3AF-3.3 As an iron core is inserted in a coil, what happens to the inductance?
A. It increases
B. It decreases
C. It stays the same
D. It becomes voltage-dependent

The answer is A. See answer 3AF-3.1.

3AF-3.4 As a brass core is inserted in a coil, what happens to the inductance?
A. It increases
B. It decreases
C. It stays the same
D. It becomes voltage-dependent

The answer is B.

3AF-3.5 For radio frequency power applications, which type of inductor has the least amount of loss?
A. Magnetic wire B. Iron core   C. Air core   D. Slug tuned

The answer is C. For radio frequency power applications, we use an inductor that has a large diameter wire and an air core. Inductors for power frequencies use an iron core. As the frequency is increased, the losses in an iron core (eddy currents and hysteresis) increase. Therefore, an air core must be used at radio frequencies.

3AF-3.6 Where does an inductor store energy?
A. In a capacitive field   B. In a magnetic field
C. In an electrical field   D. In a resistive field

The answer is B. An inductor stores energy in the form of electromagnetic lines of force that surround the inductor when current flows through it.

3AF-5.3 What is a heat sink?
A. A device used to heat an electrical component uniformly
B. A device used to remove heat from an electronic component
C. A tub in which circuit boards are soldered
D. A fan used for transmitter-cooling

The answer is B. By encasing a diode or transistor in metal and then placing the metal in intimate contact with the chassis or other good heat conductor, we form a HEAT SINK. The heat sink removes the heat from the component by conducting the heat away from the part.

SUBELEMENT 3AG
PRACTICAL CIRCUITS
(1 question)

3AG-2.1 What is a high-pass filter usually connected to?
A. The transmitter and the Transmatch
B. The Transmatch and the transmission line
C. The television receiving antenna and a television receiver's antenna input
D. The transmission line and the transmitting antenna

The answer is C. A high-pass filter is usually connected to a television set. Its purpose is to reject the lower frequency interfering signals and to pass the higher frequency television signals. See answer 3AG-2.2.

3AG-2.2 Where is the proper place to install a high-pass filter?
A. At the antenna terminals of a television receiver
B. Between a transmitter and a Transmatch
C. Between a Transmatch and the transmission line
D. On a transmitting antenna

The answer is A. A high-pass filter is installed between the TV antenna transmission line and the TV front end. It is installed as close as possible to the actual TV front end.

3AG-2.3 Where is a band-pass filter usually installed?
A. Between the spark plugs and coil in a mobile setup
B. On a transmitting antenna
C. In a communications receiver
D. Between a Transmatch and the transmitting antenna

The answer is C. A band-pass filter is installed in the IF stage of a superheterodyne receiver. We use a band-pass filter because an SSB signal or an AM signal occupies a certain amount of bandwidth, and we wish to pass the entire band of frequencies of the desired signal. The band-pass filter will reject signals outside of the desired band.
A band-pass filter can also be used between the antenna and the receiver, or any place where we wish to pass a band of frequencies and reject frequencies outside of the band.

3AG-2.4 Which frequencies are attenuated by a low-pass filter?
A. Those above its cut-off frequency
B. Those within its cut-off frequency
C. Those within 50 kHz on either side of its cut-off frequency
D. Those below its cut-off frequency

The answer is A. A low-pass filter attenuates electrical energy above its cut-off frequency. The term "low-pass" indicates that it passes low frequencies and rejects high frequencies.

3AG-2.5 What circuit passes electrical energy above a certain frequency, and attenuates electrical energy below that frequency?
A. An input filter
B. A low-pass filter
C. A high-pass filter
D. A band-pass filter

The answer is C. A high-pass filter passes electrical energy above a certain frequency, but blocks electrical energy below that frequency. The term "high-pass" indicates that high frequencies are passed and low frequencies are rejected.

3AG-2.6 What circuit passes electrical energy below a certain frequency, and blocks electrical energy above that frequency?
A. An input filter
B. A low-pass filter
C. A high-pass filter
D. A band-pass filter

The answer is B. See answer 3AG-2.4.

3AG-2.7 What circuit attenuates electrical energy above a certain frequency and below a lower frequency?
A. An input filter
B. A low-pass filter
C. A high-pass filter
D. A band-pass filter

The answer is D. A band-pass filter passes a specific band of frequencies, but rejects signals above the band and below the band. Figure 3AG-2.7 illustrates a band-pass filter.

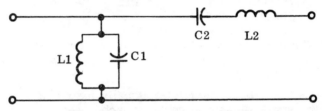

Figure 3AG-2.7. A band-pass filter.

3AG-2.9 What general range of rf energy does a band-pass filter reject?
A. All frequencies above a specified frequency
B. All frequencies below a specified frequency
C. All frequencies above the upper limit of the band in question
D. All frequencies above a specified frequency and below a lower specified frequency

The answer is D. A band-pass filter will reject signals outside of the band of frequencies it is passing. These frequencies outside of the band are undesired signals.

3AG-3.1 What circuit is likely to be found in all types of receivers?
A. A detector
B. An rf amplifier
C. An audio filter
D. A beat frequency oscillator

The answer is A. All receivers contain a detector. It is in the detector where the intelligence or desired signal is extracted from the RF energy.

3AG-3.2 In a filter-type emission J3E transmitter, what stage combines rf and af energy to produce a double-sideband suppressed carrier signal?
A. The product detector
B. The automatic-load-control circuit
C. The balanced modulator
D. The local oscillator

The answer is C. The balanced modulator combines radio frequency energy from an RF oscillator and audio frequency energy to produce a double-sideband suppressed carrier signal.

3AG-3.3 In a superheterodyne receiver for emission A3E reception, what

stage combines the received rf with energy from the local oscillator to produce a signal at the receiver intermediate frequency?
A. The mixer
B. The detector
C. The RF amplifier
D. The AF amplifier

The answer is A. The mixer is the stage that combines the incoming RF signal with the signal from the local oscillator. The mixer stage is sometimes called a first detector. This first detector is not to be confused with the second detector. It is in the second detector where the intelligence or audio is extracted from the RF.

SUBELEMENT 3AH
SIGNALS AND EMISSIONS
(2 questions)

3AH-1.1 What is emission type N0N?
A. Unmodulated carrier
B. Telegraphy by on-off keying
C. Telegraphy by keyed tone
D. Telegraphy by frequency-shift-keying
The answer is A. N0N is the steady unmodulated output of an AM transmitter.

3AH-1.2 What is emission type A3E?
A. Frequency-modulated telephony
B. Facsimile
C. Double sideband, amplitude-modulated telephony
D. Amplitude-modulated telegraphy
The answer is C. A3E is the emission of an amplitude modulated carrier. It can be speech or music.

3AH-1.3 What is emission type J3E?
A. Single-sideband suppressed-carrier amplitude-modulated telephony
B. Single-sideband suppressed-carrier amplitude-modulated telegraphy
C. Independent sideband suppressed-carrier amplitude-modulated telephony
D. Single-sideband suppressed-carrier frequency-modulated telephony
The answer is A.

3AH-1.4 What is emission type F1B?
A. Amplitude-shift-keyed telegraphy
B. Frequency-shift-keyed telegraphy
C. Frequency-modulated telephony
D. Phase-modulated telephony
The answer is B. F1B is telegraphy by frequency shift keying, without the use of a modulating audio frequency.

3AH-1.5 What is emission type F2B?
A. Frequency-modulated telephony
B. Frequency-modulated telegraphy using audio tones
C. Frequency-modulated facsimile using audio tones
D. Phase-modulated television
The answer is B. F2B is telegraphy by the on-off keying of a frequency modulating audio frequency, or by the on-off keying of a frequency modulated emission.

3AH-1.6 What is emission type F3E?
A. AM telephony   B. AM telegraphy
C. FM telegraphy   D. FM telephony
The answer is D. F3E is the emission from a transmitter that has been frequency modulated by an audio signal.

3AH-1.7 What is the emission symbol for telegraphy by frequency shift

SIGNALS AND EMISSIONS TH-2

keying without the use of a modulating tone?
A. F1B　　　B. F2B　　　C. A1A　　　D. J3E
The answer is A. See answer 3AH-1.4.

3AH-1.8 What is the emission symbol for telegraphy by the on-off keying of a frequency modulating audio tone?
A. F1B　　　B. F2A　　　C. A1A　　　D. J3E
The answer is B.

3AH-1.9 What is the emission symbol for telephony by amplitude modulation?
A. A1A　　　B. A3E　　　C. J2B　　　D. F3E
The answer is B. See answer 3AH-1.2.

3AH-1.10 What is the emission symbol for telephony by frequency modulation?
A. F2B　　　B. F3E　　　C. A3E　　　D. F1B
The answer is B. See answer 3AH-1.6.

3AH-2.2 What is the meaning of the term modulation?
A. The process of varying some characteristic of a carrier wave for the purpose of conveying information
B. The process of recovering audio information from a received signal
C. The process of increasing the average power of a single-sideband transmission
D. The process of suppressing the carrier in a single-sideband transmitter
The answer is A. See questions 3BH-2.1, 3BH-2.3, and 3BH-2.4.

3AH-6.1 What characteristic makes emission F3E especially well-suited for local VHF/UHF radio communications?
A. Good audio fidelity and intelligibility under weak-signal conditions
B. Good audio fidelity and high signal-to-noise ratio above a certain signal amplitude threshold
C. Better rejection of multipath distortion than the AM modes
D. Better carrier frequency stability than the AM modes
The answer is B. FM permits a higher signal-to-noise ratio because the noise which appears in the amplitude of the modulated signal is eliminated at the receiver. FM also exhibits the "capture effect", which means that the receiver responds only to the strongest of several signals, ignoring even a slightly weaker signal on the same channel. FM also provides a more "natural-sounding" voice reproduction than SSB.

3AH-6.2 What emission is produced by a transmitter using a reactance modulator?
A. A1A　　　B. N0N　　　C. J3E　　　D. G3E
The answer is D. G3E is phase modulated telephony. See answer 3BH-7.2.

3AH-7.1 What other emission does phase modulation most resemble?
A. Amplitude modulation　　　B. Pulse modulation
C. Frequency modulation　　　D. Single-sideband modulation
The answer is C. See answer 3BH-2.3.

3AH-9.2 Which emission does not have sidebands resulting from modulation?

A. A3E   B. N0N   C. F3E   D. F2B

The answer is B. N0N is the steady unmodulated emission of a transmitter.

3AH-12.1 To what is the deviation of an emission F3E transmission proportional?
A. Only the frequency of the audio modulating signal
B. The frequency and the amplitude of the audio modulating signal
C. The duty cycle of the audio modulating signal
D. Only the amplitude of the audio modulating signal

The answer is D. The frequency deviation of an FM emission is proportional to the amplitude of the modulating signal. Figure 3AH-12.1 illustrates a block diagram of an FM transmitter.

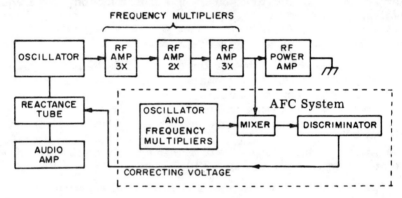

Figure 3AH-12.1. An FM transmitter.

3AH-14.1 What is the result of overdeviation of the oscillator in an emission F3E transmitter?
A. Increased transmitter power consumption
B. Out-of-channel emissions (splatter)
C. Increased transmitter range
D. Inadequate carrier suppression

The answer is B. Overdeviation in an FM transmitter results in a signal with a wider bandwidth. It generally does not distort the signal as does overmodulation in an AM transmitter.

3AH-14.2 What is splatter?
A. Interference to adjacent signals caused by excessive transmitter keying speeds.
B. Interference to adjacent signals caused by improper transmitter neutralization
C. Interference to adjacent signals caused by overmodulation of a transmitter
D. Interference to adjacent signals caused by parasitic oscillations at the antenna

The answer is C. Splatter is caused by overmodulation of a transmitter. Overmodulation results in the creation of unwanted sidebands that will interfere with signals on adjacent frequencies. These new undesirable signals are called "splatter".

SIGNALS AND EMISSIONS

3AH-16.1 What emissions are used in teleprinting?
A. F1A, F2B and F1B
B. A2B, F1B and F2B
C. A1B, A2B and F2B
D. A2B, F1A and F2B
The answer is B.

3AH-16.2 What two states of teleprinter codes are most commonly used in amateur radiocommunications?
A. Dot and dash
B. Highband and lowband
C. Start and stop
D. Mark and space
The answer is D. The mark can be used to represent the "on" condition, and the space can be used for the "off" condition.

3AH-16.3 What emission type results when an af shift keyer is connected to the microphone jack of an emission F3E transmitter?
A. A2B   B. F1B   C. F2B   D. A1F
The answer is C. F2B stands for audio frequency shift keying and since we are feeding the microphone of the transmitter with an audio frequency shift keyer, F2B emission will result.

The equipment should be able to handle the increased power resulting from this usage. Also, the user must make certain that there is no interference to other receiving equipment when using this method.

SUBELEMENT 3AI
ANTENNAS AND FEEDLINES
(3 questions)

3AI-1.1 What antenna type best strengthens signals from a particular direction while attenuating those from other directions?
A. A monopole antenna
B. An isotropic antenna
C. A vertical antenna
D. A beam antenna

The answer is D. Any type of beam or directional antenna, such as a Yagi or Quad antenna, will increase the signal reception from a desired direction. See answer 3AI-1.2.

3AI-1.2 What is a Yagi antenna?
A. Half-wavelength elements stacked vertically and excited in phase
B. Quarter-wavelength elements arranged horizontally and excited out of phase
C. Half-wavelength linear driven element(s) with parasitically excited parallel linear elements
D. Quarter-wavelength, triangular loop elements

The answer is C. A Yagi antenna consists of an ordinary half-wave dipole, plus one or more elements. Figure 3AI-1.2 illustrates a simple 3-element Yagi antenna. The half-wave dipole to which the transmission line is connected is called the driven element. The driven element receives the transmitter output power from the transmission line. The other elements are not physically connected to the transmission line or to the driven element. They are called "parasitic elements", and they receive their energy from the electromagnetic radiation of the driven element.

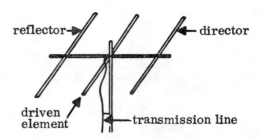

Figure 3AI-1.2. A simple 3-element Yagi antenna.

The addition of the parasitic elements increases the gain of the antenna considerably over that of a simple single element dipole. The additional antenna elements don't add RF energy to the antenna; they concentrate the radiated energy in one direction. The more the parasitic elements, the greater is the concentration or directivity of the antenna.

The transmitting advantages of a Yagi are also present in receiving. It will receive much better in the direction that it is "pointed", compared to a single element dipole. Also, it does not pick up well from other directions.

3AI-1.4 What is the general configuration of the radiating elements of a horizontally-polarized Yagi?
A. Two or more straight, parallel elements arranged in the same horizontal plane
B. Vertically stacked square or circular loops arranged in parallel horizontal planes
C. Two or more wire loops arranged in parallel vertical planes
D. A vertical radiator arranged in the center of an effective RF ground plane

The answer is A. Examine the Yagi antenna of Figure 3AI-1.2. The element in front of the driven element, in the direction of the radiation, is called the director. It is approximately 5% shorter than the driven element. The element in back of the driven element is called the reflector, and is 5% longer than the driven element. The driven element is a half-wave dipole. It is approximately one-half of the wavelength of the signal to be transmitted or received.

Regardless of the number of elements that a Yagi has, there is, in addition to the driven element, usually only one reflector; the rest of the elements are directors. Thus, a 12 element Yagi would have one driven element, one reflector, and ten directors.

3AI-1.5 What type of parasitic beam antenna uses two or more straight metal-tubing elements arranged physically parallel to each other?
A. A quad antenna
B. A delta loop antenna
C. A zepp antenna
D. A Yagi antenna

The answer is D. See answers 3AI-1.2 and 3AI-1.4.

3AI-1.6 How many directly-driven elements does a Yagi antenna have?
A. None; they are all parasitic
B. One
C. Two
D. All elements are directly driven

The answer is B. See answer 3AI-1.4.

3AI-1.8 What is a parasitic beam antenna?
A. An antenna where the director and reflector elements receive their RF excitation by induction or radiation from the driven element
B. An antenna where wave traps are used to assure magnetic coupling among the elements
C. An antenna where all elements are driven by direct connection to the feed line
D. An antenna where the driven element receives its RF excitation by induction or radiation from the directors

The answer is A. "Parasitic" is the term used to describe the elements of a Yagi antenna other than the driven element. See 3AI-1.2 and 3AI-1.4.

3AI-2.2 What kind of antenna array is composed of a square full-wave closed loop driven element with parallel parasitic element(s)?
A. Dual rhombic   B. Cubical quad   C. Stacked yagi   D. Delta loop

The answer is B. This is known as a QUAD antenna. Figure 3AI-2.2 illustrates a two-element Quad antenna.

3AI-2.3 Approximately how long is one side of the driven element of a cubical quad antenna?
A. 2 electrical wavelengths
B. 1 electrical wavelength

C. 1/2 electrical wavelength   D. 1/4 electrical wavelength
The answer is D.

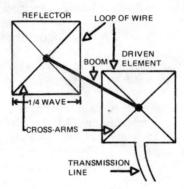

Figure 3AI-2.2. A simple 2-element quad antenna.

3AI-2.4 Approximately how long is the wire in the driven element of a cubical quad antenna?
A. 1/4 electrical wavelength   B. 1/2 electrical wavelength
C. 1 electrical wavelength   D. 2 electrical wavelengths
The answer is C.

3AI-2.5 What is a delta loop antenna?
A. A variation of the cubical quad antenna, with triangular elements
B. A large copper ring, used in direction finding
C. An antenna system composed of three vertical antennas, arranged in a triangular shape
D. An antenna made from several coils of wire on an insulating form
The answer is A. A Delta Loop is similar to a Quad antenna. However, instead of consisting of a square loop having a total length of one wavelength, it consists of three equal sides, in triangular shape, having a total length of one wavelength. See Figure 3AI-2.5.

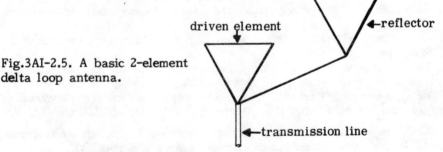

Fig.3AI-2.5. A basic 2-element delta loop antenna.

3AI-2.6 What is a cubical quad antenna?
A. Four parallel metal tubes, each approximately 1/2 electrical wavelength long
B. Two or more parallel four-sided wire loops, each approximately one electrical wavelength long
C. A vertical conductor 1/4 electrical wavelength high, fed at the bottom

ANTENNAS AND FEEDLINES

D. A center-fed wire 1/2 electrical wavelength long

The answer is B. A Quad antenna consists of one or more rectangular loops of wire. If more than one loop is used, the parasitic loops, called reflectors and/or directors, are placed in planes parallel to the driven loop. The total length of the reflector and director loops are slightly different from that of the driven loop. See Figure 3AI-2.2.

3AI-4.1 What is the polarization of electromagnetic waves radiated from a half-wavelength antenna perpendicular to the earth's surface?
A. Circularly polarized waves
B. Horizontally polarized waves
C. Parabolically polarized waves
D. Vertically polarized waves

The answer is D. An antenna that is perpendicular to the earth's surface is a vertical antenna. It produces vertically polarized waves.

3AI-4.2 What is the electromagnetic wave polarization of most man-made electrical noise radiation in the HF-VHF spectrum?
A. Left-hand circular
B. Vertical
C. Right-hand circular
D. Horizontal

The answer is B. Most man-made noise in the HF-VHF spectrum tends to be vertically polarized. Therefore, if noise is to be considered in a particular situation, we would use a horizontally polarized antenna.

3AI-4.3 To what does the term vertical as applied to wave polarization refer?
A. This means that the electric lines of force in the radio wave are parallel to the earth's surface
B. This means that the magnetic lines of force in the radio wave are perpendicular to the earth's surface
C. This means that the electric lines of force in the radio wave are perpendicular to the earth's surface
D. This means that the radio wave will leave the antenna and radiate vertically into the ionosphere

The answer is C. A radio wave consists of two perpendicular fields, one an electrical field and the other a magnetic field. The polarization of the wave is determined by the orientation of the electric field. If the electrical component of the wave is in a horizontal plane, the wave is said to be horizontally polarized. If the electrical component is perpendicular to the earth, the wave is vertically polarized. A vertical antenna produces a vertically polarized wave and a horizontal antenna produces a horizontally polarized wave.

3AI-4.4 To what does the term horizontal as applied to wave polarization refer?
A. This means that the magnetic lines of force in the radio wave are parallel to the earth's surface
B. This means that the electric lines of force in the radio wave are parallel to the earth's surface
C. This means that the electric lines of force in the radio wave are perpendicular to the earth's surface
D. This means that the radio wave will leave the antenna and radiate horizontally to the destination

The answer is B. See answer 3AI-4.3.

3AI-4.5 What electromagnetic wave polarization does a cubical quad antenna have when the feedpoint is in the center of a horizontal side?
A. Vertical   B. Horizontal   C. Circular   D. Helical
The answer is B. See Figure 3AI-2.2.

3AI-4.6 What electromagnetic wave polarization does a cubical quad antenna have when the feedpoint is in the center of a vertical side?
A. Vertical   B. Horizontal   C. Circular   D. Helical
The answer is A.

3AI-4.7 What electromagnetic wave polarization does a cubical quad antenna have when all sides are at 45 degrees to the earth's surface and the feedpoint is at the bottom corner?
A. Vertical   B. Horizontal   C. Circular   D. Helical
The answer is B.

3AI-4.8 What electromagnetic wave polarization does a cubical quad antenna have when all sides are at 45 degrees to the earth's surface and the feedpoint is at a side corner?
A. Vertical   B. Horizontal   C. Circular   D. Helical
The answer is A.

3AI-6.7 What is a directional antenna?
A. An antenna whose parasitic elements are all constructed to be directors
B. An antenna that radiates in direct line-of-sight propagation, but not skywave or skip propagation
C. An antenna permanently mounted so as to radiate in only one direction
D. An antenna that radiates more strongly in some directions than others
The answer is D.

3AI-8.1 What is meant by the term standing wave ratio?
A. The ratio of forward and reflected inductances on a feedline
B. The ratio of forward and reflected resistances on a feedline
C. The ratio of forward and reflected impedances on a feedline
D. The ratio of forward and reflected currents on a feedline
The answer is D. Standing waves are the graphs of voltage and current that occur along a resonant antenna. Standing waves occur because an RF signal traveling along an antenna, strikes the end and is reflected back. In traveling back, it either reinforces or cancels the original signal at different points. The maximums and minimums of the voltages and currents appear in the same positions along the wires, hence the name "standing waves".

The standing wave ratio (SWR) is the ratio of the maximum voltage to minimum voltage along a line. Maximum current to minimum current yields the same ratio. This is mathematically stated as follows:

$$\text{SWR} = \frac{E_{max}}{E_{min}} \text{ or } \frac{I_{max}}{I_{min}}$$

3AI-8.2 What is meant by the term forward power?
A. The power travelling from the transmitter to the antenna
B. The power radiated from the front of a directional antenna
C. The power produced during the positive half of the RF cycle

ANTENNAS AND FEEDLINES TI-6

D. The power used to drive a linear amplifier

The answer is A. Forward power is the power that leaves the transmitter and travels along the transmission line to the antenna. It can be measured by a reflectometer.

3AI-8.3 What is meant by the term reflected power?
A. The power radiated from the back of a directional antenna
B. The power returned to the transmitter from the antenna
C. The power produced during the negative half of the RF cycle
D. Power reflected to the transmitter site by buildings and trees

The answer is B. If the antenna feedpoint resistance is not equal to the characteristic impedance of the transmission line, some of the power that reaches the antenna will be reflected back to the transmitter. This is called the reflected power. Ideally, the reflected power should be equal to zero.

3AI-9.1 What is standing wave ratio a measure of?
A. The ratio of maximum to minimum voltage on a line
B. The ratio of maximum to minimum reactance on a line
C. The ratio of maximum to minimum resistance on a line
D. The ratio of maximum to minimum sidebands on a line

The answer is A. The standing wave ratio is a measure of the mismatch between the transmission line and the antenna. It causes some power to be reflected back to the transmitter from the antenna because of this mismatch. See answer 3AI-8.1.

3AI-9.2 What happens to the power loss in an unbalanced feedline as the standing wave ratio increases?
A. It is unpredictable   B. It becomes nonexistent
C. It decreases   D. It increases

The answer is D.

3AI-10.1 What is a balanced line?
A. Feed line with one conductor connected to ground
B. Feed line with both conductors connected to ground to balance out harmonics
C. Feed line with the outer conductor connected to ground at even intervals
D. Feed line with neither conductor connected to ground

The answer is D. A balanced line is a transmission line where the two conductors are alike and each conductor has the same electrical relationship to ground. See answer 3AD-14.2 and figure 3AD-14.2.

3AI-10.2 What is a balanced antenna?
A. A symmetrical antenna with one side of the feed point connected to ground
B. An antenna (or a driven element in an array) that is symmetrical about the feed point
C. A symmetrical antenna with both sides of the feed point connected to ground, to balance out harmonics
D. An antenna designed to be mounted in the center

The answer is B. The balanced antenna is a balanced system where equal voltages to ground are present at each input terminal. The common half wave dipole is an example of a balanced antenna.

3AI-10.3 What is an unbalanced line?
A. Feed line with neither conductor connected to ground
B. Feed line with both conductors connected to ground to suppress harmonics
C. Feed line with one conductor connected to ground
D. Feed line with the outer conductor connected to ground at uneven intervals

The answer is C. Coaxial cable is an example of an unbalanced line. The inner conductor is a solid or stranded wire. The outer conductor is a concentric braid which is grounded. See answer 3AD-14.3 and figure 3AD-14.3.

3AI-10.4 What is an unbalanced antenna?
A. An antenna (or a driven element in an array) that is not symmetrical about the feed point
B. A symmetrical antenna, having neither half connected to ground
C. An antenna (or a driven element in a array) that is symmetrical about the feed point
D. A symmetrical antenna with both halves coupled to ground at uneven intervals

The answer is A. A Marconi antenna is an example of an unbalanced antenna. One half of the antenna is a vertical radiator, one quarter wavelength long. The other half is represented by the ground.

3AI-11.3 What type of feedline is best suited to operating at a high standing wave ratio?
A. Coaxial cable   B. Twisted pair
C. Flat ribbon "twin lead"   D. Parallel open-wire line

The answer is D. Open-wire-parallel line can tolerate a reasonably high SWR without too much loss of power.

3AI-11.5 What is the general relationship between frequencies passing through a feed line and the losses in the feedline?
A. Loss is independent of frequency
B. Loss increases with increasing frequency
C. Loss decreases with increasing frequency
D. There is no predictable relationship

The answer is B. The feedline loss is proportional to the frequency; that is, as the frequency goes up, the losses go up, and vice versa.

3AI-11.6 What happens to rf energy not delivered to the antenna by a lossy coaxial cable?
A. It is radiated by the feedline
B. It is returned to the transmitter's chassis ground
C. Some of it is dissipated as heat in the conductors and dielectric
D. It is cancelled because of the voltage ratio of forward power to reflected power in the feedline

The answer is C. The RF energy is dissipated in the form of heat in the transmission line conductors, and in the insulation between the conductors. The RF energy is also lost through radiation from the transmission line. It is desirable that all the radiation be from the antenna.

ANTENNAS AND FEEDLINES

3AI-11.9 As the operating frequency decreases, what happens to conductor losses in a feed line?
A. The losses decrease
B. The losses increase
C. The losses remain the same
D. The losses become infinite
The answer is A.

3AI-11.11 As the operating frequency increases, what happens to conductor losses in a feed line?
A. The losses decrease
B. The losses increase
C. The losses remain the same
D. The losses decrease to zero
The answer is B.

3AI-12.3 What device can be installed on a balanced antenna so that it can be fed through a coaxial cable?
A. A triaxial transformer
B. A wavetrap
C. A loading coil
D. A balun
The answer is D. See answer 3BI-12.2.

3AI-12.4 What is a balun?
A. A device that can be used to convert an antenna designed to be fed at the center so that it may be fed at one end
B. A device that may be installed on a balanced antenna so that it may be fed with unbalanced feed line
C. A device that can be installed on an antenna to produce horizontally polarized or vertically polarized waves
D. A device used to allow an antenna to operate on more than one band
The answer is B. See answer 3BI-12.2.

SUBELEMENT 3BA
RULES AND REGULATIONS
(4 questions)

3BA-3.2 What is the maximum transmitting power permitted an amateur station on 10.14-MHz?
A. 200 watts PEP output
B. 1000 watts dc input
C. 1500 watts PEP output
D. 2000 watts dc input
The answer is A. 10.14 Mhz is in the new 30 meter band which extends from 10.1 to 10.15 MHz. This band may be used only by General, Advanced, and Extra Class operators and is limited to CW and RTTY.

3BA-3.3 What is the maximum transmitting power permitted an amateur station on 3725-kHz?
A. 200 watts PEP output
B. 1000 watts dc input
C. 1500 watts PEP output
D. 2000 watts dc input
The answer is A. This is in the Novice section of the 80 meter band.

3BA-3.4 What is the maximum transmitting power permitted an amateur station on 7080-kHz?
A. 200 watts PEP output
B. 1000 watts dc input
C. 1500 watts PEP output
D. 2000 watts dc input
The answer is C. This is in the part of the band reserved for all amateurs except Novice and Technician Class operators.

3BA-3.5 What is the maximum transmitting power permitted an amateur station on 24.95-MHz?
A. 200 watts PEP output
B. 1000 watts dc input
C. 1500 watts PEP output
D. 2000 watts dc input
The answer is C. 24.95 MHz is in the new 12 meter band that extends from 24.89 to 24.99 MHz. 24.95 MHz is in a portion of the band where CW, phone, narrow band FM and SSTV are permitted.

3BA-3.7 What is the maximum transmitting power permitted an amateur station transmitting on 21.150-MHz?
A. 200 watts PEP output
B. 1000 watts dc input
C. 1500 watts dc input
D. 1500 watts PEP output
The answer is A. This is in the Novice section of the 15 meter band.

3BA-4.1 How must a General control operator at a Novice station make the station identification when transmitting on 7050-kHz?
A. The control operator should identify the station with his or her call, followed by the word "controlling" and the Novice call
B. The control operator should identify the station with his or her call, followed by the slant bar "/" and the Novice call
C. The control operator should identify the station with the Novice call, followed by the slant bar "/" and his or her own call
D. A Novice station should not be operated on 7050 kHz, even with a General class control operator
The answer is C. 7050 kHz is NOT in the Novice part of the band. Therefore, the General Class operator must give the Novice station's call sign followed by his own call sign.

3BA-4.3 How must a newly upgraded General control operator with a Certificate of Successful Completion of Examination identify the station when transmitting on 14.325-MHz pending the receipt of a new operator license?
A. General class privileges do not include 14.325 MHz
B. No special form of identification is needed
C. The operator shall give his/her call sign, followed by the words "temporary" and the two-letter ID code shown on the certificate of successful completion
D. The operator shall give his/her call sign, followed by the date and location of the VEC examination where he/she obtained the upgraded license

The answer is C.

3BA-6.1 Under what circumstances, if any, may third-party traffic be transmitted to a foreign country by an amateur station?
A. Under no circumstances
B. Only if the country has a third-party-traffic agreement with the United States
C. Only if the control operator is an Amateur Extra class licensee
D. Only if the country has formal diplomatic relations with the United States

The answer is B.

3BA-6.2 What types of messages may be transmitted by an amateur station to a foreign country for a third-party?
A. Third-party traffic involving material compensation, either tangible or intangible, direct or indirect, to a third party, a station licensee, a control operator, or any other person
B. Third-party traffic consisting of business communications on behalf of any party
C. Only third-party traffic which does not involve material compensation of any kind, and is not business communication of any type
D. No messages may be transmitted to foreign countries for third parties

The answer is C. When transmissions between amateur stations of different countries are permitted, they shall be made in plain language and shall be limited to messages of a technical nature relating to tests and to remarks of a personal character for which, by reason of their unimportance, recourse to the public telecommunications service is not justified.

3BA-6.6 What additional limitations apply to third-party messages transmitted to foreign countries?
A. Third-party messages may only be transmitted to amateurs in countries with which the US has a third-party traffic agreement
B. Third-party messages may only be sent to amateurs in ITU Region 1
C. Third-party messages may only be sent to amateurs in ITU Region 3
D. Third-party messages must always be transmitted in English

The answer is A. It is absolutely forbidden for amateur stations to be used for transmitting international communications on behalf of third parties unless there are special arrangements between the countries concerned.

GA-3   GENERAL CLASS TEST MANUAL

3BA-8.6 Under what circumstances, if any, may an amateur station transmitting on 29.64-MHz repeat the 146.34-MHz signals of an amateur station with a Technician control operator?
A. Under no circumstances
B. Only if the station on 29.64 Mhz is operating under a Special Temporary Authorization allowing such retransmission
C. Only during an FCC-declared general state of communications emergency
D. Only if the control operator of the repeater transmitter is authorized to operate on 29.64 MHz

The answer is D. The frequencies given in the question are frequencies that are used in repeater operation. Although a Technician Class operator may not operate on 29.64 MHz, his 2 meter (146.34 MHz) signal may be retransmitted on 29.64 MHz, provided that the control operator of the 29.64 MHz transmitter has frequency privileges in this band. The important point to remember is that the control operator of the transmitting station must have the proper license.

3BA-9.1 What frequency privileges are authorized to the General operator in the 160 meter band?
A. 1800 to 1900 kHz only
B. 1900 to 2000 kHz only
C. 1800 to 2000 kHz only
D. 1825 to 2000 kHz only

The answer is C. He may operate CW or phone in the entire band.

3BA-9.2 What frequency privileges are authorized to General operators in the 75/80 meter band?
A. 3525 to 3750 and 3850 to 4000 kHz only
B. 3525 to 3775 and 3875 to 4000 kHz only
C. 3525 to 3750 and 3875 to 4000 kHz only
D. 3525 to 3775 and 3850 to 4000 kHz only

The answer is A. He may operate phone from 3.85 to 4.0 MHz. He may operate CW from 3.525 to 3.750 MHz and from 3.85 to 4.0 MHz.

3BA-9.3 What frequency privileges are authorized to the General operator in the 40 meter band?
A. 7025 to 7175 and 7200 to 7300 kHz only
B. 7025 to 7175 and 7225 to 7300 kHz only
C. 7025 to 7150 and 7200 to 7300 kHz only
D. 7025 to 7150 and 7225 to 7300 kHz only

The answer is D. He may operate CW from 7025 to 7150 kHz and from 7225 to 7300 kHz. He may operate phone from 7225 to 7300 kHz.

3BA-9.4 What frequency privileges are authorized to the General operator in the 30 meter band?
A. 10,100 to 10,150 kHz only
B. 10,105 to 10,150 kHz only
C. 10,125 to 10,150 kHz only
D. 10,100 to 10,125 kHz only

The answer is A. The actual band is from 10,100 to 10,109 kHz and 10,115 to 10,150 kHz. Only CW and RTTY are permitted.

3BA-9.5 What frequency privileges are authorized to the General operator in the 20 meter band?
A. 14,025 to 14,100 and 14,175 to 14,350 kHz only
B. 14,025 to 14,150 and 14,225 to 14,350 kHz only

RULES AND REGULATIONS

C. 14,025 to 14,125 and 14,200 to 14,350 kHz only
D. 14,025 to 14,175 and 14,250 to 14,350 kHz only
The answer is B. He may operate CW from 14.025 to 14.150 MHz. He may operate CW and phone from 14.225 to 14.350 MHz.

3BA-9.6 What frequency privileges are authorized to the General operator in the 15 meter band?
A. 21,025 to 21,200 and 21,275 to 21,450 kHz only
B. 21,025 to 21,150 and 21,300 to 21,450 kHz only
C. 21,025 to 21,200 and 21,300 to 21,450 kHz only
D. 21,000 to 21,150 and 21,275 to 21,450 kHz only
The answer is C. He may operate CW from 21.025 to 21.2 MHz and from 21.3 to 21.45 MHz. He may operate phone from 21.3 to 21.45 MHz.

3BA-9.7 What frequency privileges are authorized to the General operator in the 12 meter band?
A. 24,890 to 24,990 kHz only
B. 24,890 to 24,975 kHz only
C. 24,900 to 24,990 kHz only
D. 24,790 to 24,990 kHz only
The answer is A.

3BA-9.8 What frequency privileges are authorized to the General operator in the 10 meter band?
A. 28,000 to 29,700 kHz only
B. 28,025 to 29,700 kHz only
C. 28,100 to 29,700 kHz only
D. 28,025 to 29,600 kHz only
The answer is A. He may operate CW in the entire band. He may operate phone from 28.3 MHz to 29.7 MHz.

3BA-9.9 Which operator licenses authorize privileges on 1820-kHz?
A. Extra only
B. Extra, Advanced only
C. Extra, Advanced, General only
D. Extra, Advanced, General, Technician only
The answer is C. 1820 kHz is in the 160 meter band. All amateurs except Novice and Technician operators have privileges in this band.

3BA-9.10 Which operator licenses authorize privileges on 3950-kHz?
A. Extra, Advanced only
B. Extra, Advanced, General only
C. Extra, Advanced, General, Technician only
D. Extra, Advanced, General, Technician, Novice only
The answer is B. 3950 kHz is in that part of the 80 meter band in which Novice and Technician class operators have no privileges.

3BA-9.11 Which operator licenses authorize privileges on 7230-kHz?
A. Extra only
B. Extra, Advanced only
C. Extra, Advanced, General only
D. Extra, Advanced, General, Technician only
The answer is C. 7230 kHz is in that part of the 40 meter band in which Novice and Technician class operators have no privileges.

3BA-9.12 Which operator licenses authorize privileges on 10.125-MHz?
A. Extra, Advanced, General only
B. Extra, Advanced only
C. Extra only
D. Technician only
The answer is A. Operation in this band (30 meters) is further limited to CW and RTTY emissions on a secondary, non-interference basis with a Peak Envelope Power output not exceeding 200 watts.

3BA-9.13 Which operator licenses authorize privileges on 14.325-MHz?
A. Extra, Advanced, General, Technician only
B. Extra, Advanced, General only
C. Extra, Advanced only   D. Extra only
The answer is B. 14.325 MHz is in the 20 meter band, a band in which Novice and Technician operators have no privileges.

3BA-9.14 Which operator licenses authorize privileges on 21.425-MHz?
A. Extra, Advanced, General, Novice only
B. Extra, Advanced, General, Technician only
C. Extra, Advanced, General only   D. Extra, Advanced only
The answer is C. 21.425 MHz is in that portion of the 15 meter band where Novice and Technician class operators have no privileges.

3BA-9.15 Which operator licenses authorize privileges on 24.895-MHz?
A. Extra only   B. Extra, Advanced only
C. Extra, Advanced General only   D. None
The correct answer is C. 24.895 MHz is in the 12 meter band, a band in which Novice and Technician class operators have no privileges.

3BA-9.16 Which operator licenses authorize privileges on 29.616-MHz?
A. Novice, Technician, General, Advanced, Extra only
B. Technician, General, Advanced, Extra only
C. General, Advanced, Extra only   D. Advanced, Extra only
The answer is C. 29.616 MHz is in that portion of the 10 meter band where Novice and Technician class operators have no privileges.

3BA-10.1 On what frequencies within the 160 meter band may emission A3E be transmitted?
A. 1800-2000 kHz only   B. 1800-1900 kHz only
C. 1900-2000 kHz only   D. 1825-1950 kHz only
The answer is A. A3E stands for amplitude modulated telephony. It may be used in the entire 160 Meter band (1800-2000 kHz).

3BA-10.2 On what frequencies within the 80 meter band may emission A1A be transmitted?
A. 3500-3750 kHz only   B. 3700-3750 kHz only
C. 3500-4000 kHz only   D. 3890-4000 kHz only
The answer is C. A1A stands for CW telegraphy. It may be used in the entire 80 meter band (3500-4000 kHz).

3BA-10.3 On what frequencies within the 40 meter band may emission A3F be transmitted?
A. 7225-7300 kHz only   B. 7000-7300 kHz only
C. 7100-7150 kHz only   D. 7150-7300 kHz only
The answer is D. A3F stands for amplitude modulated television. It may be used in the 7150 to 7300 kHz part of the 40 meter band.

3BA-10.4 On what frequencies within the 30 meter band may emission F1B be transmitted?
A. 10.140-10.150 MHz only   B. 10.125-10.150 MHz only
C. 10.100-10.150 MHz only   D. 10.100-10.125 MHz only
The answer is C. F1B stands for frequency shift telegraphy without audio modulation - for automatic reception (radioteletype or RTTY). It may

RULES AND REGULATIONS GA-6

be used in the entire 30 meter band (10.1-10.15 MHz) that is available for amateur use.

3BA-10.5 On what frequencies within the 20 meter band may emission A3C be transmitted?
A. 14200-14300 kHz only
B. 14150-14350 kHz only
C. 14025-14150 kHz only
D. 14150-14300 kHz only

The answer is B. Type A3C emission stands for amplitude modulated facsimile. It may be used in the 14,150 to 14,350 kHz part of the 20 meter band.

3BA-10.6 On what frequencies within the 15 meter band may emission F3C be transmitted?
A. 21200-21300 kHz only
B. 21350-21450 kHz only
C. 21200-21450 kHz only
D. 21100-21200 kHz only

The answer is C. Type F3C emission stands for frequency modulated facsimile. It may be used in the 21,200 to 21,450 part of the 15 meter band.

3BA-10.7 On what frequencies within the 12 meter band may emission J3E be transmitted?
A. 24890-24990 kHz only
B. 24890-24930 kHz only
C. 24930-24990 kHz only
D. J3E is not permitted in this band

The answer is C. Type J3E emission stands for single sideband, suppressed carrier telephony. It may be used in the 24,930 to 24,990 kHz part of the 12 meter band.

3BA-10.8 On what frequencies within the 10 meter band may emission A3E be transmitted?
A. 28000-28300 kHz only
B. 29000-29700 kHz only
C. 28300-29700 kHz only
D. 28000-29000 kHz only

The answer is C. Type A3E emission stands for amplitude modulated telephony. It may be used in the 28.3 - 29.7 MHz part of the 10 meter band.

3BA-13.1 How is sending speed (signaling rate) for digital communications determined?
A. By taking the reciprocal of the shortest (signaling) time interval (in minutes) that occurs during a transmission, where each time interval is the period between changes of transmitter state (including changes in emission amplitude, frequency, phase, or combination of these, as authorized)
B. By taking the square root of the shortest (signaling) time interval (in seconds) that occurs during a transmission, where each time interval is the period between changes of transmitter state (including changes in emission amplitude, frequency, phase, or combination of these, as authorized)
C. By taking the reciprocal of the shortest (signaling) time interval (in seconds) that occurs during a transmission, where each time interval is the period between changes of transmitter state (including changes in emission amplitude, frequency, phase, or combination of these, as authorized)
D. By taking the square root of the shortest (signaling) time interval (in

minutes) that occurs during a transmission, where each time interval is the period between changes of transmitter state (including changes in emission amplitude, frequency, phase, or combination of these, as authorized)
The answer is C.

3BA-13.2 What is the maximum sending speed permitted for an emission F1B transmission below 28-MHz?
A. 56 kilobauds   B. 19.6 kilobauds   C. 1200 bauds   D. 300 bauds
The answer is D. Type F1B emission stands for frequency shift telegraphy without the use of a modulating audio frequency. The "B" in F1B stands for telegraphy for automatic reception (RTTY).

3BA-14.4 Under what circumstances, if any, may an amateur station engage in some form of broadcasting?
A. During severe storms, amateurs may broadcast weather information for people with scanners
B. Under no circumstances
C. If power levels under one watt are used, amateur stations may broadcast information bulletins, but not music
D. Amateur broadcasting is permissible above 10 GHz
The answer is B. An amateur station shall not be used to engage in any form of broadcasting. However, amateur stations may transmit emergency communications, information bulletins relating directly to the amateur service, round table discussions and code practice.

3BA-14.6 What protection, if any, is afforded an amateur station transmission against retransmission by a broadcast station?
A. No protection whatsoever
B. The broadcaster must secure permission for retransmission from the control operator of the amateur station
C. The broadcaster must petition the FCC for retransmission rights 30 days in advance
D. Retransmissions may only be made during a declared emergency
The answer is A. No protection is afforded an amateur station. According to a recent FCC Public Notice, broadcast stations are free to broadcast amateur radio transmissions with or without the approval of the amateur operators involved. However, the broadcasters may not become actively involved in the amateur transmission.

3BA-15.1 Under what circumstances, if any, may the playing of a violin be transmitted by an amateur station?
A. When the music played produces no dissonances or spurious emissions
B. When it is used to jam an illegal transmission
C. Only above 1215 MHz
D. Transmitting music is not permitted in the Amateur Service
The answer is D. Playing a violin is music and the transmission of music by an amateur station is prohibited.

3BA-15.3 Under what circumstances, if any, may the playing of a piano be transmitted by an amateur station?
A. When it is used to jam an illegal transmission
B. Only above 1215 MHz

C. Transmitting music is not permitted in the Amateur Service
D. When the music played produces no dissonances or spurious emissions
The answer is C. Playing a piano is music and the transmission of music by an amateur station is prohibited.

3BA-15.4 Under what circumstances, if any, may the playing of a harmonica be transmitted by an amateur station?
A. When the music played produces no dissonances or spurious emissions
B. Transmitting music is not permitted in the Amateur Service
C. When it is used to jam an illegal transmission
D. Only above 1215 MHz
The answer is B. Harmonica playing is considered music and the transmission of music by an amateur station is prohibited.

3BA-16.1 Under what circumstances, if any, may an amateur station transmit a message in a secret code in order to obscure the meaning?
A. Only above 450 MHz   B. Only on Field Day
C. Never   D. Only during a declared communications emergency
The answer is C.

3BA-16.2 What types of abbreviations or signals are not considered codes or ciphers?
A. Abbreviations and signals certified by the ARRL
B. Abbreviations and signals established by regulation or custom and usage and whose intent is to facilitate communication and not to obscure meaning
C. No abbreviations are permitted, as they tend to obscure the meaning of the message to FCC monitoring stations
D. Only "10-codes" are permitted
The answer is B.

3BA-16.3 When, if ever, are codes and ciphers permitted in domestic amateur radiocommunications?
A. Codes and ciphers are prohibited under all circumstances
B. Codes and ciphers are permitted during ARRL-sponsored contests
C. Codes and ciphers are permitted during nationally declared emergencies
D. Codes and ciphers are permitted above 2.3 GHz
The answer is A.

3BA-16.4 When, if ever, are codes and ciphers permitted in international amateur radiocommunications?
A. Codes and ciphers are prohibited under all circumstances
B. Codes and ciphers are permitted during ITU-sponsored DX contests
C. Codes and ciphers are permitted during internationally declared emergencies
D. Codes and ciphers are permitted only on frequencies above 2.3 GHz
The answer is A.

SUBELEMENT 3BB
OPERATING PROCEEDURES
(3 questions)

3BB-1.4 What is meant by the term flattopping in an emission J3E transmission?
A. Signal distortion caused by insufficient collector current
B. The transmitter's automatic level control is properly adjusted
C. Signal distortion caused by excessive drive
D. The transmitter's carrier is properly suppressed

The answer is C. J3E stands for single sideband, suppressed carrier telephony. Figure 3BD-3.3B illustrates a signal that has "flattopping". This comes about when too strong a signal (excessive drive) is fed to the grid of the final RF amplifier tube.

A signal with flattopping produces severe distortion and adjacent channel interference, called splatter. One way to avoid flattopping is to reduce the audio peaks, thereby reducing the drive to the final stage.

3BB-1.5 How should the microphone gain control be adjusted on an emission J3E transmitter?
A. For full deflection of the ALC meter on modulation peaks
B. For slight movement of the ALC meter on modulation peaks
C. For 100% frequency deviation on modulation peaks
D. For a dip in plate current

The answer is B. J3E stands for single sideband, suppressed carrier telephony. Adjusting the microphone gain control prevents distortion by automatically limiting the gain of prior stages. However, the audio gain should be high enough for proper modulation.

3BB-2.1 In which segment of the 20 meter band do most emission F1B transmissions take place?
A. Between 14.000 and 14.050 MHz.
B. Between 14.075 and 14.100 MHz.
C. Between 14.150 and 14.225 MHz.
D. Between 14.275 and 14.350 MHz.

The answer is B. F1B is the emission designator for frequency shift telegraphy for automatic reception (RTTY). RTTY is legally permitted between 14.000 and 14.150 MHz; however, by agreement among amateurs, it is confined to 14.075-14.100 MHz.

3BB-2.2 In which segment of the 80 meter band do most emission F1B transmissions take place?
A. 3.610 to 3.630 MHz.   B. 3500 to 3525 kHz.
C. 3700 to 3750 kHz.   D. 3.775 to 3.825 MHz.

The answer is A. F1B is the emission designator for frequency shift telegraphy for automatic reception (RTTY). RTTY is legally permitted between 3.5 and 3.75 MHz; however, by agreement among amateurs, it is confined to 3.61-3.63 MHz.

3BB-2.3 What is meant by the term Baudot?

A. Baudot is a 7-bit code, with start, stop and parity bits
B. Baudot is a 7-bit code in which each character has four mark and three space bits
C. Baudot is a 5-bit code, with additional start and stop bits
D. Baudot is a 6-bit code, with additional start, stop and parity bits

The answer is C. Baudot is a code used in radioteletype. In the Baudot code, each character is made up of five data elements plus a start element and a stop element. An element is either a space(0) or a mark(1). The different characters (letters of the alphabet, numbers, etc.) have different combinations of spaces and marks. Figure 3BB-2.3A shows the letter "J". The five data pulses that distinguish the letter J are: mark, mark, space, mark, space. Figure 3BB-2.3B illustrates the letter "Y". The five data pulses for Y are mark, space, mark, space, mark. Before the five data pulses, there is a start pulse which is always a space pulse. After the five data pulses there is a stop pulse which is always a mark pulse.

When the speed is 60 words per minute, the time duration for each data pulse and start pulse is 22 ms (milliseconds); the duration for the stop pulse is 31 ms. Higher code speeds have shorter duration pulses.

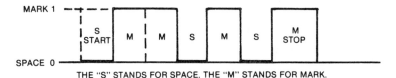

Fig. 3BB-2.3A. The letter "J" in Baudot code.

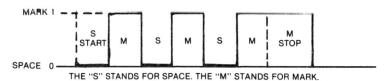

Fig. 3BB-2.3B. The letter "Y" in Baudot code.

3BB-2.4 What is meant by the term ASCII?
A. ASCII is a 7-bit code, with additional start, stop and parity bits
B. ASCII is a 7-bit code in which each character has four mark and three space bits
C. ASCII is a 5-bit code, with additional start and stop bits
D. ASCII is a 5-bit code in which each character has three mark and two space bits

The answer is A. ASCII is the abbreviation for American Standard Code for Information Interchange. It was adopted in 1968. Whereas the Baudot Code contains five data bits, ASCII has seven data bits. The seven data bits permit many more combinations (128) for the ASCII than the five data bits for the Baudot Code. The ASCII code has separate combinations for upper and lower case letters; the Baudot Code does not because it has only 32 combinations. An eighth bit may be added to the ASCII Code for error detection. It is called a parity bit.

3BB-2.6 What is the most common frequency shift for emission F1B transmissions in the amateur HF bands?
A. 85 Hz.   B. 170 Hz.   C. 425 Hz.   D. 850 Hz.

The answer is B. F1B is the emission designator for frequency shift telegraphy (RTTY). 850 Hz. was originally used as the frequency shift. However, today, the frequency shift of 170 Hz. is almost always used.

3BB-2.10 What are the two subset modes of AMTOR?
A. A mark of 2125 Hz. and a space of 2295 Hz.
B. Baudot and ASCII   C. ARQ and FEC   D. USB and LSB

The answer is C. See answer 3AB-2.5. ARQ stands for Automatic Repeat Request. It is also called Mode A. FEC stands for Forward Error Corrections. It is also called Mode B. In the ARQ mode, the two stations continuously check each other and the signal is repeated only when it is requested. In the FEC mode each character is sent out twice.

3BB-2.11 What is the meaning of the term ARQ?
A. Automatic Repeater Queue
B. Automatic Receiver Quieting
C. Automatically Resend Quickly
D. Automatic Repeat Request

The answer is D. See answer to question 3BB-2.10.

3BB-2.12 What is the meaning of the term FEC?
A. Frame Error Check
B. Forward Error Correction
C. Frequency Envelope Control
D. Frequency Encoded Connection

The answer is B. See answer to question 3BB-2.10.

3BB-3.8 What is a bandplan?
A. An outline adopted by Amateur Radio operators for operating within a specific portion of radio spectrum
B. An arrangement for deviating from FCC Rules and Regulations
C. A schedule for operating devised by the Federal Communications Commission
D. A plan devised for a club on how best to use a band during a contest

The answer is A. A bandplan is a plan for allocating frequency segments in a band for specific types of operating. This allows for orderly operating practice with minimum interference. Bandplans are developed by amateur organizations and are not a part of the official FCC rules. However, the FCC does encourage bandplans and considers them to be in accordance with good amateur practice.

3BB-3.12 What is the usual input/output frequency separation for a 10 meter station in repeater operation?
A. 100 kHz   B. 600 kHz   C. 1.6 MHz   D. 170 Hz

The answer is A. See answer 3AB-3.9.

3BB-4.1 What is meant by the term VOX transmitter control?
A. Circuitry that causes the transmitter to transmit automatically when the operator speaks into the microphone
B. Circuitry that shifts the frequency of the transmitter when the opera-

OPERATING PROCEDURES

GB-4

tor switches from radiotelegraphy to radiotelephony
C. Circuitry that activates the receiver incremental tuning in a transceiver
D. Circuitry that isolates the microphone from the ambient noise level

The answer is A. VOX stands for VOICE OPERATED TRANSMISSION. The VOX circuit controls the transmit-receive changeover circuit. When the operator speaks into the microphone, the transmitter is turned on and the receiver off. When the operator ceases talking, the transmitter is turned off and the receiver is turned on.

3BB-4.2 What is the common name for the circuit that causes a transmitter to automatically transmit when a person speaks into the microphone?
A. VXO   B. VOX   C. VCO   D. VFO

The answer is B. See answer 3BB-4.1.

3BB-5.1 What is meant by the term full break-in telegraphy?
A. A system of radiotelegraph communication in which the breaking station sends the Morse Code symbol BK
B. A system of radiotelegraph communication in which only automatic keyers can be used
C. A system of radiotelegraph communication in which the operator must activate the send-receive switch after completing a transmission
D. A system of radiotelegraph communication in which the receiver is sensitive to incoming signals between transmitted key pulses

The answer is D. Full break-in telegraphy is a system where an operator who is transmitting can hear the other operator between his sending of dots and dashes. This allows for more efficient operating.

3BB-5.2 What Q signal is used to indicate full break-in telegraphy capability?
A. QSB   B. QSF   C. QSK   D. QSV

The answer is C.

3BB-6.1 When selecting an emission A1A transmitting frequency, what is the minimum frequency separation from a QSO in progress that should be allowed in order to minimize interference?
A. 5 to 50 Hz   B. 150 to 500 Hz
C. Approximately 3 kHz   D. Approximately 6 kHz

The answer is B. A1A stands for CW. While it is true that CW is merely an interruption of a single frequency carrier, a CW signal does occupy a certain amount of frequency spectrum. The dit-dah interruption of a carrier does produce a limited degree of modulation, which results in a limited bandwidth. This bandwidth is determined by the speed of sending and the shape of the keyed signal. The bandwidth of a CW signal is far less than that of a phone signal. It is in the order of a few hundred hertz.

3BB-6.2 When selecting an emission J3E transmitting frequency, what is the minimum frequency separation from a QSO in progress that should be allowed in order to minimize interference?
A. 150 to 500 Hz between suppressed carriers
B. Approximately 3 kHz between suppressed carriers
C. Approximately 6 kHz between suppressed carriers

D. Approximately 10 kHz between suppressed carriers

The answer is B. J3E stands for single sideband, suppressed carrier telephony. The audio of a phone signal is limited to between 2.5 and 3 kHz. Therefore, your frequency should be 3 kHz away from an adjacent signal.

3BB-6.3 When selecting an emission F1B RTTY transmitting frequency, what is the minimum frequency separation from a QSO in progress that should be allowed in order to minimize interference?
A. Approximately 45 Hz center to center
B. Approximately 250 to 500 Hz center to center
C. Approximately 3 kHz center to center
D. Approximately 6 kHz center to center

The answer is B. F1B means frequency-shift keying (fsk). When F1B is used, the transmitter carrier is shifted between two frequencies. The shift or frequency difference between the two frequencies is commonly 170 Hz. Thus, we must allow a bandwidth of at least this amount and therefore, stay away from an adjacent signal by at least a few hundred Hertz.

3BB-7.1 What is an azimuthal map?
A. A map projection that is always centered on the North Pole
B. A map projection, centered on a particular location, that determines the shortest path between two points on the surface of the earth
C. A map that shows the angle at which an amateur satellite crosses the equator
D. A map that shows the number of degrees longitude that an amateur satellite appears to move westward at the equator with each orbit

The answer is B. It is the best kind of map to use when trying to orient a directional antenna for finding the shortest path to a DX station.

3BB-7.2 How can an azimuthal map be helpful in conducting international HF radiocommunications?
A. It is used to determine the proper beam heading for the shortest path to a DX station
B. It is used to determine the most efficient transmitting antenna height to conduct the desired communication
C. It is used to determine the angle at which an amateur satellite crosses the equator
D. It is used to determine the maximum usable frequency (MUF)

The answer is A. See answer 3BB-7.1.

3BB-7.3 What is the most useful type of map when orienting a directional antenna toward a station 5,000 miles distant?
A. Azimuthal   B. Mercator
C. Polar projection   D. Topographical

The answer is A. See answer 3BB-7.1.

3BB-7.4 A directional antenna pointed in the long-path direction to another station is generally oriented how many degrees from the short-path heading?
A. 45 degrees   B. 90 degrees   C. 180 degrees   D. 270 degrees

The answer is C. Since the long path is diametrically opposite to the short-path, the antenna would have to be pointed 180 degrees away from the short-path.

3BB-7.5 What is the short-path heading to Antarctica?
A. Approximately 0 degrees
B. Approximately 90 degrees
C. Approximately 180 degrees
D. Approximately 270 degrees
The answer is C. In navigation, azimuth is the angular distance from the North point, measured Eastward and around the circle in a horizontal plane. Since Antarctica is directly South, the azimuth is 180 degrees and the short-path beam should be 180 degrees.

3BB-8.1 When permitted, transmissions to amateur stations in another country must be limited to only what type of messages?
A. Messages of any type are permitted
B. Messages that compete with public telecommunications services
C. Messages of a technical nature or remarks of a personal character of relative unimportance
D. Such transmissions are never permitted
The answer is C. Business messages or political discussions should be avoided.

3BB-8.2 In which International Telecommunication Union Region is the continental United States?
A. Region 1   B. Region 2   C. Region 3   D. Region 4
The answer is B. The I.T.U is the world organization that allocates frequency use to the various services. The I.T.U. has, for purposes of frequency allocation, divided the world into three regions. Region 1 consists principally of Europe and Africa. Region 2 includes North and South America. Region 3 consists primarily of Southern Asia and Australia.

3BB-8.3 In which International Telecommunication Union Region is Alaska?
A. Region 1   B. Region 2   C. Region 3   D. Region 4
The answer is B. See answer 3BB-8.2.

3BB-8.4 In which International Telecommunication Union Region is American Samoa?
A. Region 1   B. Region 2   C. Region 3   D. Region 4
The answer is C. See answer 3BB-8.2.

3BB-8.5 For uniformity in international radiocommunication, what time measurement standard should amateur radio operators worldwide use?
A. Eastern Standard Time   B. Uniform Calibrated Time
C. Universal Coordinated Time D. Universal Time Control
The answer is C. Universal Coordinated Time is the standard time that is used in International events, and is also used by radio amateurs in dealing with other amateurs in different time zones. It is the local time in Greenwich, England, and is also referred to as Greenwich Mean Time (GMT). Universal Coordinated Time is abbreviated UTC from the French "Universal Temps Coordonne".

3BB-8.6 In which International Telecommunications Union Region is Hawaii?
A. Region 1   B. Region 2   C. Region 3   D. Region 4
The answer is B. See answer to question 3BB-8.2.

3BB-8.7 In which International Telecommunications Union Region is the Commonwealth of Northern Marianas Islands?
A. Region 1   B. Region 2   C. Region 3   D. Region 4
   The answer is C. See answer to question 3BB-8.2.

3BB-8.8 In which International Telecommunications Union Region is Guam?
A. Region 1   B. Region 2   C. Region 3   D. Region 4
   The answer is C. See answer to question 3BB-8.2

3BB-8.9 In which International Telecommunications Union Region is Wake Island?
A. Region 1   B. Region 2   C. Region 3   D. Region 4
   The answer is C. See answer to question 3BB-8.2.

3BB-10.1 What is the Amateur Auxiliary to the FCC's Field Operations Bureau?
A. Amateur Volunteers formally enlisted to monitor the airwaves for rules violations
B. Amateur Volunteers who conduct Amateur Radio licensing examinations
C. Amateur Volunteers who conduct frequency coordination for amateur VHF repeaters
D. Amateur Volunteers who determine height above average terrain measurements for repeater installations
   The answer is A. The Communications Amendments Act of 1982 (Public Law 97-259) authorizes the FCC to use amateur volunteers to help monitor the airwaves for rules violations. The FCC's Field Operations Bureau has created an Amateur Auxiliary which is to be administered by the ARRL.

3BB-10.2 What are the objectives of the Amateur Auxiliary to the FCC's Field Operations Bureau?
A. To enforce amateur self-regulation and compliance with the rules
B. To foster amateur self-regulation and compliance with the rules
C. To promote efficient and orderly spectrum usage in the repeater sub-bands
D. To provide emergency and public safety communications
   The answer is B. The primary purpose of the Amateur Auxiliary is to provide help to other amateurs and to find the causes of amateur problems. It is NOT to ENFORCE the rules.

SUBELEMENT 3BC
RADIO WAVE PROPAGATION
(3 questions)

3BC-1.6 What is the maximum distance along the earth's surface that can normally be covered in one hop using the F2 layer?
A. Approximately 180 miles
B. Approximately 1200 miles
C. Approximately 2500 miles
D. No distance. This layer does not support radio communication
The answer is C.

3BC-1.7 What is the maximum distance along the earth's surface that can be covered in one hop using the E layer?
A. Approximately 180 miles
B. Approximately 1200 miles
C. Approximately 2500 miles
D. No distance. This layer does not support radio communication
The answer is B.

3BC-1.9 What is the average height of maximum ionization of the E layer?
A. 45 miles   B. 70 miles   C. 200 miles   D. 1200 miles
The answer is B. See Figure 3AC-1.1.

3BC-1.10 During what part of the day, and in what season of the year can the F2 layer be expected to reach its maximum height?
A. At noon during the summer   B. At midnight during the summer
C. At dusk in the spring and fall   D. At noon during the winter
The answer is A. However, the F layer can rise quite high at night.

3BC-1.13 What is the critical angle, as used in radio wave propagation?
A. The lowest take off angle that will return a radio wave to earth under specific ionospheric conditions
B. The compass direction of the desired DX station from your location
C. The 180-degree-inverted compass direction of the desired DX station from your location
D. The highest take off angle that will return a radio wave to earth during specific ionospheric conditions
The answer is D. The angle of radiation is the angle between the sky wave and the earth's surface. This is shown in Figure 3BC-1.13. It can be seen that as the angle of radiation is reduced, the skywave is refracted back to a more distant point on earth. However, as the angle of radiation is increased, the sky wave penetrates the ionosphere and is NOT returned to earth. The angle above which there is no return of the signal to earth and below which signals are returned to earth, is called the "critical angle". The value of the critical angle depends on the frequency and other ionospheric conditions.

3BC-2.3 What is the main reason that the 160, 80 and 40 meter amateur bands tend to be useful for only short-distance communications during daylight hours?

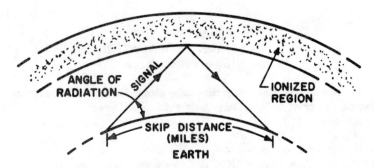

Figure 3BC-1.13. Propagation of sky waves.

A. Because of a lack of activity
B. Because of auroral propagation
C. Because of D-layer absorption
D. Because of magnetic flux

The answer is C. The main reason is absorption by the D layer which limits the range of communications for these bands. See answers 3AC-1.2 and 3AC-1.8.

3BC-2.4 What is the principal reason the 160 meter through 40 meter bands are useful for only short-distance radiocommunications during daylight hours?
A. F-layer bending   B. Gamma radiation
C. D-layer absorption   D. Tropospheric ducting

The answer is C. See answers 3AC-1.2 and 3AC-1.8.

3BC-3.3 If the maximum usable frequency on the path from Minnesota to Africa is 22-MHz, which band should offer the best chance for a successful QSO?
A. 10 meters   B. 15 meters   C. 20 meters   D. 40 meters

The answer is B. We use the next lowest amateur frequency band, which is 21.0 to 21.45 MHz. This is the 15 meter band.

3BC-3.4 If the maximum usable frequency on the path from Ohio to West Germany is 17-MHz, which band should offer the best chance for a successful QSO?
A. 80 meters   B. 40 meters   C. 20 meters   D. 2 meters

The answer is C. We use the next lowest amateur frequency band, which is 14.0 to 14.35 MHz. This is the 20 meter band.

3BC-5.1 Over what periods of time do sudden ionospheric disturbances normally last?
A. The entire day
B. A few minutes to a few hours
C. A few hours to a few days
D. Approximately one week

The answer is B. A Sudden Ionospheric Disturbance, abbreviated SID, is the result of a sudden flare-up or eruption on the sun. It causes strong radiation from the sun that ionizes the D layer of the ionosphere. This absorbs signals in the HF range and results in an almost total absence of sky wave reception. This fade-out lasts from a few minutes to a few hours.

3BC-5.2 What can be done at an amateur station to continue radiocommunications during a sudden ionospheric disturbance?
A. Try a different frequency
B. Try the other sideband
C. Try a different antenna polarization
D. Try a different frequency shift
   The answer is A. Since different frequencies are affected differently by the ionosphere, the only thing that we can do is to try a different band, preferably a higher frequency band. Sudden Ionospheric Disturbances affect the lower frequencies more than the higher frequencies.

3BC-5.3 What effect does a sudden ionospheric disturbance have on the daylight ionospheric propagation of HF radio waves?
A. Disrupts higher-latitude paths more than lower-latitude paths
B. Disrupts transmissions on lower frequencies more than those on higher frequencies
C. Disrupts communications via satellite more than direct communications
D. None. Only dark (as in nighttime) areas of the globe are affected
   The answer is B. See answers 3BC-5.1 and 3BC-5.2.

3BC-5.4 How long does it take a solar disturbance that increases the sun's ultraviolet radiation to cause ionospheric disturbances on earth?
A. Instantaneously   B. 1.5 seconds
C. 8 minutes   D. 20 to 40 hours
   The answer is C. It takes about 8 minutes for the effects of ultraviolet radiation to show up in a disturbance of radio waves. This is because ultraviolet radiation travels at about the speed of light, 186,000 miles per second, and the sun is 92,900,000 miles away from the earth.

3BC-5.5 Sudden ionospheric disturbances cause increased radio wave absorption in which layer of the ionosphere?
A. D layer   B. E layer   C. F1 layer   D. F2 layer
   The answer is A. See answer 3BC-5.1.

3BC-6.2 What is a characteristic of backscatter signals?
A. High intelligibility   B. A wavering sound
C. Reversed modulation   D. Reversed sidebands
   The answer is B. Backscatter signals are generally weak and distorted. They have a watery or echo type of sound. See answer 3AC-6.1.

3BC-6.4 What makes backscatter signals often sound distorted?
A. Auroral activity and changes in the earth's magnetic field
B. The propagation through ground waves that absorb much of the signal's clarity
C. The earth's E-layer at the point of radio wave refraction
D. The small part of the signal's energy scattered back to the transmitter skip zone through several radio-wave paths
   The answer is D. This is because of multipath effects. The signal arrives at a given point via several paths. Since the signals arriving from several paths are not in phase, distortion will occur. See answer 3AC-6.1.

3BC-6.5 What is the radio wave propagation phenomenon that allows a

signal to be detected at a distance too far for ground wave propagation but too near for normal sky wave propagation?
A. Ground wave
B. Scatter
C. Sporadic-E skip
D. Short path skip

The answer is B. More specifically, it would be backscatter. See answer 3AC-6.1.

3BC-6.6 When does ionospheric scatter propagation on the HF bands most often occur?
A. When the sunspot cycle is at a minimum
B. At night
C. When the F1 and F2 layers are combined
D. At frequencies above the maximum usable frequency

The answer is D.

3BC-7.1 What is solar flux?
A. The density of the sun's magnetic field
B. The radio energy emitted by the sun
C. The number of sun spots on the side of the sun facing the earth
D. A measure of the tilt of the earth's ionosphere on the side toward the sun

The answer is B. The sun emits ultraviolet rays and particles that cause ionization of the ionosphere. We refer to the sun's radiated energy as solar flux. The solar flux can be measured by special receivers.

3BC-7.2 What is the solar-flux index?
A. A measure of past measurements of solar activity
B. A measurement of solar activity that compares daily readings with results from the last six months
C. Another name for the American sun spot number
D. A measure of solar activity that is taken daily

The answer is D. The radiation from the sun can be picked up as noise on a receiver. The amount of noise picked up depends upon the amount of solar activity and varies considerably. The unit used to measure and compare solar radiation is called the "solar flux index". The solar flux index is measured daily at 1700 (5 PM) UTC at an observatory in Ottawa, Canada. This measurement is made on a frequency of 2800 MHz. The readings are broadcast by the National Bureau of Standards radio station WWV.

3BC-7.3 What is a timely indicator of solar activity?
A. The 2800-MHz solar flux index
B. The mean Canadian sunspot number
C. A clock set to Coordinated Universal Time
D. Van Allen radiation measurements taken at Boulder Colorado

The answer is A. See answer 3BC-7.2.

3BC-7.4 What type of propagation conditions on the 15 meter band is indicated by a solar-flux index value of 60 to 70?
A. Unpredictable ionospheric propagation
B. No ionospheric propagation is possible
C. Excellent ionospheric propagation
D. Poor ionospheric propagation

The answer is D. An index above 85 is needed for good activity on the 15 meter band. See answer 3BC-7.2.

3BC-7.5 A solar flux index in the range of 90 to 110 indicates what type of propagation conditions on the 15 meter band?
A. Poor ionospheric propagation
B. No ionospheric propagation is possible
C. Unpredictable ionospheric propagation
D. Good ionospheric propagation
The answer is D. See answers to questions 3BC-7.2 and 3BC-7.4.

3BC-7.6 A solar flux index of greater than 120 would indicate what type of propagation conditions on the 10 meter band?
A. Good ionospheric propagation
B. Poor ionospheric propagation
C. No ionospheric propagation is possible
D. Unpredictable ionospheric propagation
The answer is A. An index of 120 is more than adequate for the 10 meter band. See answer 3BC-7.2.

3BC-7.7 For widespread long distance openings on the 6 meter band, what solar-flux values would be required?
A. Less than 50   B. Approximately 75
C. Approximately 100   D. Greater than 250
The answer is D. The 6 meter band is not normally a good band for long distance communications, so it would take a high solar flux of at least 200 to make long distance communications possible.

3BC-7.8 If the MUF is high and HF radiocommunications are generally good for several days, a similar condition can usually be expected how many days later?
A. 7 days   B. 14 days   C. 28 days   D. 90 days
The answer is C. The MUF and HF radiocommunications are due to solar activity which correlates with the sunspots on the sun's surface. Since the sun's rotation cycle is approximately 28 days, it will take that amount of time for the same sunspots to reappear with its consequent similar propagation conditions.

3BC-10.1 What is a geomagnetic disturbance?
A. A sudden drop in the solar-flux index
B. A shifting of the earth's magnetic pole
C. Ripples in the ionosphere
D. A dramatic change in the earth's magnetic field over a short period of time
The answer is D. The earth itself is a magnet with magnetic poles and a magnetic field. The earth's magnetic field affects radio wave propagation. Strong radiation of particles from the sun will cause disturbances in the earth's magnetic field, which in turn, will affect radio wave propagation. These disturbances in the earth's magnetic field are called "geomagnetic disturbances".

3BC-10.2 Which latitude paths are more susceptible to geomagnetic disturbances?

A. Those greater than 45 degrees latitude
B. Those less than 45 degrees latitude
C. Equatorial paths
D. All paths are affected equally
  The answer is A.

3BC-10.3 What can be the effect of a major geomagnetic storm on radiocommunications?
A. Improved high-latitude HF communications
B. Degraded high-latitude HF communications
C. Improved ground-wave propagation
D. Improved chances of ducting at UHF
  The answer is B. See answer 3BC-10.1.

3BC-10.4 How long does it take a solar disturbance that increases the sun's radiation of charged particles to affect radio wave propagation on earth?
A. The effect is instantaneous   B. 1.5 seconds
C. 8 minutes   D. 20 to 40 hours
  The answer is D. It takes from one to two days for the effect to show up in a disturbance of radio wave propagation. This is because charged particles travel "slowly" compared to the speed of light. See answer to question 3BC-5.4.

SUBELEMENT 3BD
AMATEUR RADIO PRACTICE
(5 questions)

3BD-1.5 Which wires in a four conductor line cord should be attached to fuses in a 234-vac primary (single-phase) power supply?
A. Only the "hot" (black and red) wires
B. Only the "neutral" (white) wire
C. Only the ground (bare) wire
D. All wires

The answer is A. This is extremely important. Only the hot leads should be fused. If the other leads were fused and the fuse opened up, there would be a serious shock hazard.

3BD-1.6 What size wire is normally used on a 15-ampere, 117-vac household lighting circuit?
A. AWG No.14   B. AWG No.16   C. AWG No.18   D. AWG No.22

The answer is A. This is arrived at by consulting a table that lists the current-carrying capacity of the various sizes of wire. If in doubt, always use a larger size of wire for safety and minimum voltage drop. The smaller AWG numbers indicate a larger diameter wire. The more current that has to be carried, the greater should be the diameter of the wire.

3BD-1.7 What size wire is normally used on a 20-ampere, 117-vac household appliance circuit?
A. AWG No.20   B. AWG No.16   C. AWG No.14   D. AWG No.12

The answer is D. See answer 3BD-1.6. Note that a 20 ampere circuit requires a smaller NUMBER AWG wire size than a 15 ampere circuit.

3BD-1.8 What could be a cause of the room lights dimming when the transmitter is keyed?
A. RF in the ac pole transformer
B. High resistance in the key contacts
C. A drop in ac line voltage
D. The line cord is wired incorrectly

The answer is C. When the transmitter is keyed, a high current is drawn from the power line. From Ohm's Law, we know that the voltage drop along the line will increase ($E = I \times R$). This will leave less voltage available at the room light socket and the lights will dim. An example will illustrate this condition. Assume that the power source voltage is 117 volts and the voltage drop along the line is 2 volts. This will leave 115 volts at the transmitter and light sockets. If the transmitter should then be keyed, more current would be drawn through the line and the voltage drop along the line would increase, let us say, from 2 volts to 6 volts. This would leave 111 volts at the transmitter and light sockets (117V - 6V = 111V), and the room lights will dim.

3BD-1.9 What size fuse should be used on a #12 wire household appliance circuit?
A. Maximum of 100 amperes   B. Maximum of 60 amperes

C. Maximum of 30 amperes   D. Maximum of 20 amperes

The answer is D. Since #12 wire should only carry 20 amperes, the fuse should be no greater than 20 amperes. If we use a larger fuse, it will not limit the current in the line to its proper safe capacity.

3BD-2.4 What safety feature is provided by a bleeder resistor in a power supply?
A. It improves voltage regulation
B. It discharges the filter capacitors
C. It removes shock hazards from the induction coils
D. It eliminates ground-loop current

The answer is B. The primary purpose of a bleeder resistor is to discharge the filter capacitors when the power supply is shut down. The bleeder resistor is also used to improve voltage regulation and to tap reduced voltages where required.

3BD-3.1 What kind of input signal is used to test the amplitude linearity of an emission J3E transmitter while viewing the output on an oscilloscope?
A. Normal speech
B. An audio frequency sine wave
C. Two audio frequency sine waves
D. An audio frequency square wave

The answer is C. Emission J3E stands for single sideband, suppressed carrier telephony. A two-tone test signal is used. The two tones are two audio frequency signals of equal amplitude and approximately 1 kHz apart in frequency. Each of the two tones must be free of harmonics.

3BD-3.2 To test the amplitude linearity of an emission J3E transmitter with an oscilloscope, what should the audio input to the transmitter be?
A. Normal speech
B. An audio frequency sine wave
C. Two audio frequency sine waves
D. An audio frequency square wave

The answer is C. It should be two audio frequency sine wave signals. See answer 3BD-3.1.

3BD-3.3 How are two-tones used to test the amplitude linearity of an emission J3E transmitter?
A. Two harmonically-related audio tones are fed into the microphone input of a J3E transmitter, and the output is observed on an oscilloscope
B. Two harmonically-related audio tones are fed into the microphone input of the transmitter, and the output is observed on a distortion analyzer
C. Two non-harmonically related audio tones are fed into the microphone input of the transmitter, and the output is observed on an oscilloscope
D. Two non-harmonically related audio tones are fed into the microphone input of the transmitter, and the output is observed on a wattmeter

The answer is C. The two audio tones are mixed together and fed to the transmitter microphone input. A spectrum analyzer is used to show the transmitter output on an oscilloscope. The amplitude of the test signal output, along with that of the distortion products, can be seen and measured on the oscilloscope.

AMATEUR RADIO PRACTICE

Figure 3BD-3.3 illustrates oscilloscope patterns using a two-tone test signal. A normal pattern is shown at A. B shows a flattened pattern due to overdrive. C shows a pattern resulting from excessive bias.

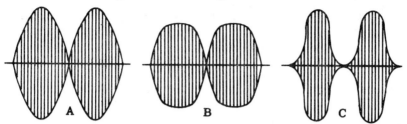

Figure 3BD-3.3. Oscilloscope patterns using a two-tone test.

3BD-3.4 What audio frequencies are used in a two-tone test of the linearity of an emission J3E transmitter?
A. 20 Hz and 20,000 Hz tones must be used
B. 1200 Hz and 2400 Hz tones must be used
C. Any two audio tones may be used, if they are harmonically related
D. Any two audio tones may be used, but they must be within the transmitter audio passband, and should not be harmonically related

The answer is D. The two tones must be sine wave, non-harmonically related tones. They should be of equal amplitude, free of harmonics and approximately 1,000 Hz apart. The two tones should be low enough in frequency to fit inside the audio pass band of the SSB transmitter.

3BD-3.5 What can be determined by making a two-tone test using an oscilloscope?
A. The percent of frequency modulation
B. The percent of carrier phase shift
C. The frequency deviation
D. The amplifier linearity

The answer is D. The purpose of the two-tone test is primarily to test and analyze the SSB transmitter for various types of distortion, including the signal-to-distortion ratio. It may also be used to check for hum and carrier balance. The oscilloscope is used to observe the signal at the output of the transmitter.

3BD-4.1 How can the grid-current meter in a power amplifier be used as a neutralizing indicator?
A. Tune for minimum change in grid current as the output circuit is changed
B. Tune for maximum change in grid current as the output circuit is changed
C. Tune for minimum grid current
D. Tune for maximum grid current

The answer is A. The grid current meter is inserted into the grid circuit of the stage to be neutralized. The plate tank capacitor is rocked back and forth while observing the meter. The neutralizing capacitor is adjusted for minimum change in the meter reading while rocking the plate tank capacitor.

3BD-4.2 Why is neutralization in some vacuum tube amplifiers necessary?
A. To reduce the limits of loaded Q in practical tuned circuits
B. To reduce grid to cathode leakage
C. To cancel acid build-up caused by thorium oxide gas
D. To cancel oscillation caused by the effects of interelectrode capacitance

The answer is D. An RF stage is neutralized in order to prevent it from oscillating and emitting spurious signals. It oscillates because of the feedback from the plate circuit to the grid circuit, through the plate-grid capacitance of the tube. When the RF stage is designed, it is neutralized by adding a small capacitor in such a way that it neutralizes or cancels out the existing interelectrode capacity, thereby preventing the stage from oscillating.

3BD-4.3 How is neutralization of an rf amplifier accomplished?
A. By supplying energy from the amplifier output to the input on alternate half cycles
B. By supplying energy from the amplifier output to the input shifted 360 degrees out of phase
C. By supplying energy from the amplifier output to the input shifted 180 degrees out of phase
D. By supplying energy from the amplifier output to the input with a proper DC bias

The answer is C. A common operating procedure for neutralizing an RF amplifier is as follows:

(1) Remove the plate voltage from the stage to be neutralized. This is very important since it is impossible to neutralize an amplifier with the plate voltage applied.

(2) The stage preceding the stage to be neutralized should have its power on and should be properly tuned. The filament voltage of the stage being neutralized should be on.

(3) The grid and plate tank circuits of the stage being neutralized should now be tuned to resonate with the signal coming from the preceding stage. This is done by tuning the grid and plate tank capacitors for a maximum indication of the RF indicator, which is coupled to the plate tank circuit. The RF indicator may be a neon bulb or a flashlight bulb, connected to a small loop of wire. We can also use a thermocouple ammeter which is connected in series with the plate tank circuit. If a thermocouple is used, care should be taken not to overload the meter. It is a sensitive instrument that is easily damaged. Any coupling between an indicating instrument and the plate tank coil should be made as loose as possible.

(4) The neutralizing capacitor or capacitors are then adjusted until the RF indicator shows that the RF energy in the plate tank circuit is at a minimum.

(5) Repeat steps 3 and 4 to make sure that the stage is as completely neutralized as possible.

3BD-4.4 What purpose does a neutralizing circuit serve in an rf amplifier?
A. It controls differential gain
B. It cancels the effects of positive feedback

C. It eliminates circulating currents
D. It reduces incidental grid modulation
The answer is B. See answer 3BD-4.2.

3BD-4.5 What is the reason for neutralizing the final amplifier stage of a transmitter?
A. To limit the modulation index
B. To eliminate parasitic oscillations
C. To cut off the final amplifier during standby periods
D. To keep the carrier on frequency
The answer is B. See answer 3BD-4.2.

3BD-5.1 How can the output PEP of a transmitter be determined with an oscilloscope?
A. Measure peak load voltage across a resistive load with an oscilloscope, and calculate, using PEP = [(Vp)(Vp)]/(Rl)
B. Measure peak load voltage across a resistive load with an oscilloscope, and calculate, using PEP = [(0.707 PEV)(0.707 PEV)]/Rl
C. Measure peak load voltage across a resistive load with an oscilloscope, and calculate, using PEP = (Vp)(Vp)(Rl)
D. Measure peak load voltage across a resistive load with an oscilloscope and calculate, using PEP = [(1.414 PEV)(1.414 PEV)]/Rl
The answer is B. It is a simple matter to measure the Peak Envelope Voltage on the oscilloscope screen. We then use the basic formula, P = E x E/R, to calculate the Peak Envelope Power. We multiply by .707 because, despite an apparent contradiction, the Peak Envelope Power output of a modulated transmitter is an average value over an RF cycle rather than an absolute Peak value.

3BD-5.5 What is the output PEP from a transmitter when an oscilloscope shows 200-volts peak-to-peak across a 50 ohm resistor connected to the transmitter output terminals?
A. 100 watts   B. 200 watts   C. 400 watts   D. 1000 watts
The answer is A. The formula used to solve this problem is:

$$\text{PEP output power} = \frac{(\text{Peak Voltage} \times 0.707)^2}{R_{load}}$$

This formula is derived from the basic formula, P = E x E/R. The peak voltage is one half of the peak-to-peak voltage. See figure 3BE-16.2. The peak voltage is therefore 100 volts. We then substitute the known values in the PEP output power formula above and solve the equation.

$$\text{PEP output power} = \frac{(100 \times 0.707)^2}{50} = \frac{70.7^2}{50} = 100 \text{ watts}$$

3BD-5.6 What is the output PEP from a transmitter when an oscilloscope shows 500-volts peak-to-peak across a 50 ohm resistor connected to the transmitter output terminals?
A. 500 watts   B. 625 watts   C. 1250 watts   D. 2500 watts

The answer is B. See answer to question 3BD-5.5.

$$\text{PEP output power} = \frac{(250 \times .707)^2}{50} = \frac{176.75^2}{50} = 625 \text{ watts}$$

3BD-5.7 What is the output PEP from an N0N transmitter when an average-reading wattmeter connected to the transmitter output terminals indicates 1060 watts?
A. 530 watts   B. 1060 watts   C. 1500 watts   D. 2120 watts

The answer is B. N0N stands for a steady, unmodulated carrier emission. The PEP output power is therefore the same as the average power.

3BD-6.1 What item of test equipment contains horizontal and vertical channel amplifiers?
A. The ohmmeter   B. The signal generator
C. The ammeter   D. The oscilloscope

The answer is D. The horizontal amplifier amplifies the signal that is going to the horizontal plates of the oscilloscope, and the vertical amplifier amplifies the signal going to the vertical plates.

3BD-6.2 What types of signals does an oscilloscope measure?
A. Any time-dependent signal within the bandwidth capability of the instrument
B. Blinker-light signals from ocean-going vessels
C. International nautical flag signals
D. Signals created by aeronautical flares

The answer is A. An oscilloscope is an electrical instrument that displays, on a screen, the wave shape of a signal. The signal to be displayed can be DC or AC. The frequency of the signal to be displayed can be quite high, depending upon the oscilloscope.

3BD-6.3 What is an oscilloscope?
A. An instrument that displays the radiation resistance of an antenna
B. An instrument that displays the SWR on a feed line
C. An instrument that displays the resistance in a circuit
D. An instrument that displays signal waveforms

The answer is D. See answer 3BD-6.2.

3BD-6.4 What can cause phosphor damage to an oscilloscope cathode ray tube?
A. Directly connecting deflection electrodes to the cathode ray tube
B. Too high an intensity setting
C. Overdriving the vertical amplifier
D. Improperly adjusted focus

The answer is B. We can damage the screen of an oscilloscope CRT if we leave a dot or line on the screen for a long period of time, especially if the intensity is set too high.

3BD-9.1 What is a signal tracer?
A. A direction-finding antenna
B. An aid for following schematic diagrams
C. A device for detecting signals in a circuit
D. A device for drawing signal waveforms

The answer is C. A signal tracer consists of a detector and an indicating device (speaker, meter, etc.), connected to an input probe. The signal tracer is applied to different parts or stages of a circuit and indicates whether or not a signal is present and its intensity. In this way, it can identify an inoperative stage. The signal being traced is usually supplied by a signal generator or signal injector. The signal tracer is useful for trouble shooting defective equipment in an amateur station, particularly receivers.

3BD-9.2 How is a signal tracer used?
A. To detect the presence of a signal in the various stages of a receiver
B. To locate a source of interference
C. To trace the path of a radio signal through the ionosphere
D. To draw a wave form on paper
The answer is A. See answer 3BD-9.1.

3BD-9.3 What is a signal tracer normally used for?
A. To identify the source of radio transmissions
B. To make exact replicas of signals
C. To give a visual indication of standing waves on open-wire feed lines
D. To identify an inoperative stage in a radio receiver
The answer is D. See answer 3BD-9.1.

3BD-10.1 What is the most effective way to reduce or eliminate audio frequency interference to home entertainment systems?
A. Install bypass inductors
B. Install bypass capacitors
C. Install metal oxide varistors
D. Install bypass resistors
The answer is B. A strong RF signal will be rectified and detected in an early stage of improperly designed audio equipment. This interference can be minimized only by modification of the audio equipment. Some of the things that can be done are: Shielding and/or by-passing of the power leads, speaker leads, and other interconnecting leads. Also, the grid of the first audio tube should be by-passed with a .001 mfd. capacitor.

3BD-10.2 What should be done when a properly-operating amateur station is the source of interference to a nearby telephone?
A. Make internal adjustments to the telephone equipment
B. Contact a phone service representative about installing RFI filters
C. Nothing can be done to cure the interference
D. Ground and shield the local telephone distribution amplifier
The answer is B. The amateur operator should suggest that his neighbor notify the telephone company. The telephone company can easily cure the problem by by-passing the telephone microphone and lines with RF by-pass capacitors.

3BD-10.3 What sound is heard from a public address system when audio rectification occurs in response to a nearby emission J3E transmission?
A. A steady hum that persists while the transmitter's carrier is on the air
B. On-and-off humming or clicking
C. Distorted speech from the transmitter's signals
D. Clearly audible speech from the transmitter's signals

The answer is C. The simple diode detection that occurs in a public address system is not entirely adequate in detecting an SSB signal. An SSB receiver has a special Product detector. The amateur SSB signal will therefore sound distorted in the output of the public address system.

3BD-10.4 How can the possibility of audio rectification occurring be minimized?
A. By using a solid state transmitter
B. By using CW emission only
C. By ensuring all station equipment is properly grounded
D. By using AM emission only

The answer is C. There is little that the amateur can do at his station to reduce this type of interference, other than cut his power down considerably, and make certain that his equipment is properly grounded. The fault lies with the audio equipment. This interference can be minimized by modification of the audio equipment. Some of the things that can be done are: shielding and/or by-passing of the power leads, speaker leads, and other interconnecting leads. Also, the grid of the first audio tube should be by-passed with a .001 mfd. capacitor.

From a practical point of view, it is not a good idea to work on your neighbor's equipment. You should suggest that they contact the manufacturer of the equipment. The manufacturer may provide parts or information so that a local serviceman can modify the equipment to prevent interference to it.

3BD-10.5 What sound is heard from a public address system when audio rectification occurs in response to a nearby emission A3E transmission?
A. Audible, possibly distorted speech from the transmitter signals
B. On-and-off humming or clicking
C. Muffled, distorted speech from the transmitter's signals
D. Extremely loud, severely distorted speech from the transmitter's signals

The answer is A. See answer 3BD-10.3. An AM signal can be properly detected in the simple diode detection that occurs in a public address system. The audio will therefore sound clear.

3BD-12.2 What is the reason for using a speech processor with an emission J3E transmitter?
A. A properly-adjusted speech processor reduces average transmitter power requirements
B. A properly-adjusted speech processor reduces unwanted noise pickup from the microphone
C. A properly-adjusted speech processor improves voice frequency fidelity
D. A properly-adjusted speech processor improves signal intelligibility at the receiver

The answer is D. A speech processor reduces the ratio between signal peaks and the average power level when used with an SSB transmitter. This increases the average effective radiated power, which results in improved speech intelligibility at the receiver, especially under poor transmission conditions.

3BD-12.3 When a transmitter is 100% modulated, will a speech processor increase the output PEP?

A. Yes   B. No
C. It will decrease the transmitter's peak power output
D. It will decrease the transmitter's average power output
    The answer is B. It increases the average power. See answer 3BD-12.2.

3BD-12.4 Under which band conditions should a speech processor not be used?
A. When there is high atmospheric noise on the band
B. When the band is crowded
C. When the frequency in use is clear
D. When the sunspot count is relatively high
    The answer is C. It is not required when we have a clear channel and reception is excellent.

3BD-12.5 What effect can result from using a speech processor with an emission J3E transmitter?
A. A properly-adjusted speech processor reduces average transmitter power requirements
B. A properly-adjusted speech processor reduces unwanted noise pickup from the microphone
C. A properly-adjusted speech processor improves voice frequency fidelity
D. A properly-adjusted speech processor improves signal intelligibility at the receiver
    The answer is D. It will increase the average power output of the transmitter, thereby increasing the intelligibility of the signal at the receiver.

3BD-13.1 At what point in a coaxial line should an electronic T-R switch be installed?
A. Between the transmitter and low-pass filter
B. Between the low-pass filter and antenna
C. At the antenna feedpoint
D. Right after the low-pass filter
    The answer is A. One disadvantage of a T-R switch is that it may generate harmonics which will cause TVI or other interference. Therefore, it is a good idea to use a low-pass filter between the T-R switch and the antenna to eliminate these harmonics. Figure 3BD-13.1 shows where the T-R switch is placed in the amateur station.

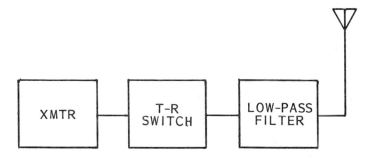

Figure 3BD-13.1. Placement of a T-R switch.

3BD-13.2 Why is an electronic T-R switch preferable to a mechanical one?
A. Greater receiver sensitivity
B. Circuit simplicity
C. Higher operating speed
D. Cleaner output signals

The answer is C. An electronic T-R switch allows for much faster operating than a manual switch. It also permits the station to be used with full break-in capability on CW and VOX on speech.

3BD-13.3 What station accessory facilitates QSK operation?
A. Oscilloscope
B. Audio CW filter
C. Antenna relay
D. Electronic T/R switch

The answer is D. QSK means full break-in telegraphy. See answer 3BD-13.2.

3BD-14.6 What is an antenna noise bridge?
A. An instrument for measuring the noise figure of an antenna or other electrical circuit
B. An instrument for measuring the impedance of an antenna or other electrical circuit
C. An instrument for measuring solar flux
D. An instrument for tuning out noise in a receiver

The answer is B. An antenna noise bridge is an instrument that measures the impedance of an antenna. The noise bridge consists of a wideband noise generator and an RF bridge circuit. A zener diode is used as the noise source. The antenna is connected to the instrument so that it becomes one arm of the bridge. A known reactance and resistance in the bridge become another arm of the bridge. When the RF noise generator is turned on, the known reactance and resistance are adjusted for null on the detector (which is the receiver). By looking at the readings of the known values at null, we can easily calculate the antenna impedance.

3BD-14.7 How is an antenna noise bridge used?
A. It is connected at the antenna feed point, and the noise is read directly
B. It is connected between a transmitter and an antenna and tuned for minimum SWR
C. It is connected between a receiver and an unknown impedance and tuned for minimum noise
D. It is connected between an antenna and a Transmatch and adjusted for minimum SWR

The answer is C. One jack on the noise bridge is connected to the antenna. Another jack on the noise bridge is connected to the receiver. The noise source is turned on and the bridge is adjusted for minimum (null) noise. See answer 3BD-14.6.

3BD-15.1 How does the emitted waveform from a properly adjusted emission J3E transmitter appear on a monitoring oscilloscope?
A. A vertical line
B. A waveform that mirrors the input waveform
C. A square wave
D. Two loops at right angles

The answer is B. The monitoring oscilloscope is used to check the

AMATEUR RADIO PRACTICE GD-11

quality of the emitted signal. By examining the waveform, we can spot distortion and other problems in the output signal.

3BD-15.2 What is the best instrument for checking transmitted signal quality from an emission A1A/J3E transmitter?
A. A monitor oscilloscope
B. A field strength meter
C. A sidetone monitor
D. A diode probe and an audio amplifier
   The answer is A. See answer 3BD-15.1.

3BD-15.3 What is a monitoring oscilloscope?
A. A device used by the FCC to detect out-of-band signals
B. A device used to observe the waveform of a transmitted signal
C. A device used to display SSTV signals
D. A device used to display signals in a receiver IF stage
   The answer is B. The monitoring scope can be on all the time, thereby continually observing (monitoring) the output signal.

3BD-15.4 How is a monitoring oscilloscope connected in a station in order to check the quality of the transmitted signal?
A. Connect the receiver IF output to the vertical-deflection plates of the oscilloscope
B. Connect the transmitter audio input to the oscilloscope vertical input
C. Connect a receiving antenna directly to the oscilloscope vertical input
D. Connect the transmitter output to the vertical-deflection plates of the oscilloscope
   The answer is D. When we are observing a signal on an oscilloscope and if we want to see the amplitude of the signal waveform versus time, we apply the signal to the vertical plates of the oscilloscope. The horizontal plates are normally connected to the internal sweep of the oscilloscope.

3BD-17.2 What is the most appropriate instrument to use when determining antenna horizontal radiation patterns?
A. A field strength meter   B. A grid-dip meter
C. A wave meter   D. A vacuum-tube voltmeter
   The answer is A. A field strength meter is a simple but useful piece of test equipment that is used to tune and adjust antennas, to determine the approximate directivity of beams, or as a simple "confidence check" to confirm transmitter power output at any time.
   A basic field strength meter consists of a tuned circuit, a silicon diode, and a DC milliammeter. It does not need a power supply and uses only a short antenna (either a telescoping whip or a random length of wire). The field strength meter detects RF power and indicates relative signal strength on the milliammeter.

3BD-17.3 What is a field-strength meter?
A. A device for determining the standing-wave ratio on a transmission line
B. A device for checking modulation on the output of a transmitter
C. A device for monitoring relative RF output
D. A device for increasing the average transmitter output
   The answer is C. See answer 3BD-17.2.

3BD-17.4 What is a simple instrument that can be useful for monitoring relative rf output during antenna and transmitter adjustments?
A. A field-strength meter
B. An antenna noise bridge
C. A multimeter
D. A Transmatch

The answer is A. See answer 3BD-17.2.

3BD-17.5 When the power output from a transmitter is increased by four times, how should the S-meter reading on a nearby receiver change?
A. Decrease by approximately one S-unit
B. Increase by approximately one S-unit
C. Increase by approximately four S-units
D. Decrease by approximately four S-units

The answer is B. Increasing the power by a factor of four is a 6 dB power increase. This is derived from the dB power formula:

$$dB = 10 \log \frac{P2}{P1} = 10 \log 4 = 10 \times .6021 = 6 \text{ dB}$$

(the log table tells us that the log of 4 is 0.6021)

Each S unit on an S-meter is equivalent to approximately 6 dB. Therefore, a power increase of four times results in an increase in the S-meter of one S-unit.

3BD-17.6 By how many times must the power output from a transmitter be increased to raise the S-meter reading on a nearby receiver from S-8 to S-9?
A. Approximately 2 times
B. Approximately 3 times
C. Approximately 4 times
D. Approximately 5 times

The answer is C. This question is the exact opposite of question 3BD-17.5. If in question 3BD-17.5, a power increase of 4 times causes an S-meter increase of one S-unit, simple logic tells us that a one S-unit increase on the S-meter is caused by a power increase of four times.

This can also be proven by substituting in the dB power formula. We put 6 dB on the left side of the formula because we know that 6 dB is the equivalent of one S-unit.

$$6 \text{ dB} = 10 \log X, \quad .6 \text{ dB} = \log X$$

The log table tells us that the log of 4 is equal to .6. Therefore, the power increases four times.

SUBELEMENT 3BE
ELECTRICAL PRINCIPLES
(2 questions)

3BE-1.1 What is meant by the term impedance?
A. The electric charge stored by a capacitor
B. The opposition to the flow of ac in a circuit containing only capacitance
C. The opposition to the flow of ac in a circuit
D. The force of repulsion presented to an electric field by another field with the same charge

The answer is C. The impedance is the TOTAL opposition to the flow of alternating current. It is used in a circuit containing both resistance and reactance. The formula for the impedance of an AC circuit containing both reactance and resistance is:

$$\text{Impedance (z)} = \sqrt{R^2 + X^2}$$

3BE-1.2 What is the opposition to the flow of ac in a circuit containing both resistance and reactance called?
A. Ohm   B. Joule   C. Impedance   D. Watt

The answer is C. See answer 3BE-1.1.

3BE-3.1 What is meant by the term reactance?
A. Opposition to dc caused by resistors
B. Opposition to ac caused by inductors and capacitors
C. A property of ideal resistors in ac circuits
D. A large spark produced at switch contacts when an inductor is de-energized

The answer is B. Reactance is the opposition to AC by coils and capacitors. Inductive reactance is the opposition of an inductance to AC. Capacitive reactance is the opposition of a capacitor to AC.

3BE-3.2 What is the opposition to the flow of ac caused by an inductor called?
A. Resistance   B. Reluctance   C. Admittance   D. Reactance

The answer is D. See answer 3BE-3.1.

3BE-3.3 What is the opposition to the flow of ac caused by a capacitor called?
A. Resistance   B. Reluctance   C. Admittance   D. Reactance

The answer is D. See answer 3BE-3.1.

3BE-3.4 How does a coil react to ac?
A. As the frequency of the applied ac increases, the reactance decreases
B. As the amplitude of the applied ac increases, the reactance also increases
C. As the amplitude of the applied ac increases, the reactance decreases
D. As the frequency of the applied ac increases, the reactance also increases

The answer is D. A coil is an inductor. Inductive reactance varies directly with frequency. That is, as the frequency increases, the inductive reactance increases; as the frequency decreases, the inductive reactance decreases.

3BE-3.5 How does a capacitor react to ac?
A. As the frequency of the applied ac increases, the reactance decreases
B. As the frequency of the applied ac increases, the reactance increases
C. As the amplitude of the applied ac increases, the reactance also increases
D. As the amplitude of the applied ac increases, the reactance decreases

The answer is A. Capacitive reactance varies inversely with frequency. That is, as the frequency increases, the capacitive reactance decreases; as the frequency decreases, the capacitive reactance increases.

3BE-6.1 When will a power source deliver maximum output?
A. When the impedance of the load is equal to the impedance of the source
B. When the SWR has reached a maximum value
C. When the power supply fuse rating equals the primary winding current
D. When air wound transformers are used instead of iron core transformers

The answer is A.

3BE-6.2 What is meant by impedance matching?
A. To make the load impedance much greater than the source impedance
B. To make the load impedance much less than the source impedance
C. To use a balun at the antenna feed point
D. To make the load impedance equal the source impedance

The answer is D. When energy is being transferred from a source to a load, maximum energy will be transferred when the impedance of the load is equal to the impedance of the source.

3BE-6.3 What occurs when the impedance of an electrical load is equal to the internal impedance of the power source?
A. The source delivers minimum power to the load
B. There will be a high SWR condition
C. No current can flow through the circuit
D. The source delivers maximum power to the load

The answer is D. See answer 3BE-6.2.

3BE-6.4 Why is impedance matching important in radio work?
A. So the source can deliver maximum power to the load
B. So the load will draw minimum power from the source
C. To ensure that there is less resistance than reactance in the circuit
D. To ensure that the resistance and reactance in the circuit are equal

The answer is A. See answer 3BE-6.2.

3BE-7.2 What is the unit measurement of reactance?
A. Mho B. Ohm C. Ampere D. Siemen

The answer is B. The ohm is the unit of all forms of opposition in a circuit.

3BE-7.4 What is the unit measurement of impedance?
A. Ohm B. Volt C. Ampere D. Watt

ELECTRICAL PRINCIPLES

The answer is A. See answers 3AE-7.1 and 3BE-7.2.

3BE-10.1 What is a bel?
A. The basic unit used to describe a change in power levels
B. The basic unit used to describe a change in inductances
C. The basic unit used to describe a change in capacitances
D. The basic unit used to describe a change in resistances

The answer is A. The **bel** (abbreviated B) is a unit used to express a ratio between two power, current, or voltage levels in sound and electrical work. The bel is a logarithmic unit. This is because our impression of loudness is proportional to the logarithm of the increase in sound energy and not to the increase of the energy itself. For example, if a sound were increased in energy to 1,000 times its original value, it would only appear to the ear to be 30 times as loud. The **decibel** (abbreviated dB), which is equal to one tenth of a bel, is more commonly used.

3BE-10.2 What is a decibel?
A. A unit used to describe a change in power levels, equal to 0.1 bel
B. A unit used to describe a change in power levels, equal to 0.01 bel
C. A unit used to describe a change in power levels, equal to 10 bels
D. A unit used to describe a change in power levels, equal to 100 bels

The answer is A. See answer 3BE-10.1.

3BE-10.3 Under ideal conditions, a barely detectable change in loudness is approximately how many dB?
A. 12 dB   B. 6 dB   C. 3 dB   D. 1 dB

The answer is D. See answer 3BE-10.1.

3BE-10.4 A two-times increase in power results in a change of how many dB?
A. Multiplying the original power by 2 gives a new power that is 1 dB higher
B. Multiplying the original power by 2 gives a new power that is 3 dB higher
C. Multiplying the original power by 2 gives a new power that is 6 dB higher
D. Multiplying the original power by 2 gives a new power that is 12 dB higher

The answer is B. We can determine this from the power formula, substituting 0.301 for log 2.

$$dB = 10 \log \frac{P2}{P1} = 10 \log 2 = 10 \times 0.301 = 3 \text{ dB}$$

Thus we see that multiplying the power by 2 gives a 3 dB increase.

3BE-10.5 An increase of 6dB results from raising the power by how many times?
A. Multiply the original power by 1.5 to get the new power
B. Multiply the original power by 2 to get the new power
C. Multiply the original power by 3 to get the new power
D. Multiply the original power by 4 to get the new power

The answer is D. We use the formula given in 3BE-10.4

$$6 = 10 \log x \qquad 0.6 = \log x \qquad 0.6 = \log 4$$

The log table tells us that the log of 4 is equal to 0.6. Therefore, the power increase is 4 times.

3BE-10.6 A decrease of 3dB results from lowering the power by how many times?
A. Divide the original power by 1.5 to get the new power
B. Divide the original power by 2 to get the new power
C. Divide the original power by 3 to get the new power
D. Divide the original power by 4 to get the new power

The answer is B. This question is similar to question 3BE-10.4. If we multiply power by 2, we have an increase of 3 dB. Similarly, if we divide the power by 2 we have a decrease of 3 dB.

3BE-10.7 A signal strength report is "10dB over S9". If the transmitter power is reduced from 1500 watts to 150 watts, what should be the new signal strength report?
A. S5   B. S7   C. S9   D. S9 plus 5 dB

The answer is C. We substitute the powers given in the question into the dB power formula given in answer 3BE-10.4.

$$dB = 10 \log \frac{1500}{150} = 10 \log 10$$

The log of 10 is 1. Therefore, dB = 10 X 1 = 10 dB.
The reduction would be 10 dB. The new signal strength report would be "S9".

3BE-10.8 A signal strength report is "20dB over S9". If the transmitter power is reduced from 1500 watts to 150 watts, what should be the new signal strength report?
A. S5   B. S7   C. S9   D. S9 plus 10 dB

The answer is D. This problem is solved in the same manner as question 3BE-10.7. Since we start with 20 dB over S9 a reduction of 10 dB leaves us with 10db over S9.

3BE-10.9 A signal strength report is "20dB over S9". If the transmitter power is reduced from 1500 watts to 15 watts, what should be the new signal strength report?
A. S5   B. S7   C. S9   D. S9 plus 10 dB

The answer is C. We substitute the powers given in the question into the dB power formula given in answer 3BE-10.4.

$$dB = 10 \log \frac{1500}{15} = 10 \log 100$$

The log of 100 is 2. Therefore, dB = 10 x 2 = 20 dB.
The reduction would be 20 dB. The new signal strength report would therefore be "S9".

**3BE-12.1 If a 1.0 ampere current source is connected to two parallel-

connected 10 ohm resistors, how much current passes through each resistor?
A. 10 amperes   B. 2 amperes   C. 1 ampere   D. 0.5 ampere
   The answer is D. In a parallel connected circuit, the sum of the individual branch currents is equal to the source current. Since the two resistors are equal in value, equal currents flow through each of them. Therefore, one-half ampere flows through each resistor. This is shown in Figure 3BE-12.1.

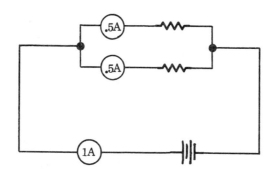

Figure 3BE-12.1. Two resistors in parallel

3BE-12.3 In a parallel circuit with a voltage source and several branch resistors, what relationship does the total current have to the current in the branch circuits?
A. The total current equals the average of the branch currents through each resistor
B. The total current equals the sum of the branch currents through each resistor
C. The total current decreases as more parallel resistors are added to the circuit
D. The total current is calculated by adding the voltage drops across each resistor and multiplying the sum by the total number of all circuit resistors
   The answer is B. See answer 3BE-12.1.

3BE-13.1 How many watts of electrical power are being consumed when a 400-vdc power source supplies an 800 ohm load?
A. 0.5 watt is dissipated   B. 200 watts are dissipated
C. 400 watts are dissipated   D. 320000 watts are dissipated
   The answer is B. We use a basic power formula to find the power dissipated.

$$P = \frac{E^2}{R} = \frac{400 \times 400}{800} = \frac{160,000}{800} = 200 \text{ watts}$$

3BE-13.2 How many watts of electrical power are being consumed by a 12-vdc pilot light which draws 0.2-amperes?
A. 60 watts   B. 24 watts   C. 6 watts   D. 2.4 watts
   The answer is D. In order to find power when we know the voltage and

current, we use a basic power formula:

$$P = E \times I = 12 \times 0.2 = 2.4 \text{ watts}$$

It should be noted that, technically speaking, power is not dissipated or "consumed". The heat ENERGY is dissipated and POWER is the RATE at which electrical energy is converted into heat energy.

3BE-13.3 How many watts are being dissipated when 7.0-milliamperes flow through 1.25-kilohms?
A. Approximately 61 milliwatts B. Approximately 39 milliwatts
C. Approximately 11 milliwatts D. Approximately 9 milliwatts

The answer is A. The three formulas used to solve problems involving the power, current and resistance are:

$$P = I^2 \times R \qquad R = \frac{P}{I^2} \qquad I = \frac{P}{R}$$

where: P is power in watts,
I is current in amperes,
R is the resistance in ohms,

In this problem, we use the first formula. However, we must change milliamperes to amperes and kilohms to ohms before we substitute in the formula.

$$7.0 \text{ ma.} = 0.007 \text{ A.} \quad 1.25 \text{ kilohms} = 1250 \text{ ohms}$$

$$P = I^2 \times R = 0.007 \times 0.007 \times 1250$$

$$P = 0.061 \text{ watts or 61 milliwatts.}$$

3BE-14.1 How is the total resistance calculated for several resistors in series?
A. The total resistance must be divided by the number of resistors to ensure accurate measurement of resistance
B. The total resistance is always the lowest-rated resistance
C. The total resistance is found by adding the individual resistances together
D. The tolerance of each resistor must be raised proportionally to the number of resistors

The answer is C.

3BE-14.2 What is the total resistance of two equal, parallel-connected resistors?
A. Twice the resistance of either resistance
B. The sum of the two resistances
C. The total resistance cannot be determined without knowing the exact resistances
D. Half the resistance of either resistor

The answer is D. The formula for determining the total resistance of ANY TWO resistors in parallel is as follows:

$$R_{total} = \frac{R1 \times R2}{R1 + R2}$$

If the two resistors in parallel are equal, then the total resistance is always equal to one half of the value of either resistor.

ELECTRICAL PRINCIPLES

3BE-14.3 What is the total inductance of two equal, parallel connected inductors?
A. Half the inductance of either inductor, assuming no mutual coupling
B. Twice the inductance of either inductor, assuming no mutual coupling
C. The sum of the two inductances, assuming no mutual coupling
D. The total inductance cannot be determined without knowing the exact inductances

The answer is A. Inductors in parallel behave in the same manner as resistors in parallel. Therefore, the total inductance of two inductors connected in parallel is equal to the product of their individual inductances divided by the sum of their individual inductances. See 3BE-14.2.

3BE-14.4 What is the total capacitance of two equal, parallel-connected capacitors?
A. Half the capacitance of either capacitor
B. Twice the capacitance of either capacitor
C. The value of either capacitor
D. The total capacitance cannot be determined without knowing the exact capacitances

The answer is B. The total capacity of any number of capacitors in parallel is equal to the simple sum of the capacitances. Since the two capacitances are equal in this question, the total capacitance is twice the value of either capacitor.

3BE-14.5 What is the total resistance of two equal, series-connected resistors?
A. Half the resistance of either resistor
B. Twice the resistance of either resistance
C. The value of either resistor
D. The total resistance cannot be determined without knowing the exact resistances

The answer is B. The total resistance of any number of resistors connected in series is the simple sum of all the resistances. Since the two resistors in this problem are equal, the total resistance is twice the resistance value of either resistor.

3BE-14.6 What is the total inductance of two equal, series-connected inductors?
A. Half the inductance of either inductor, assuming no mutual coupling
B. Twice the inductance of either inductor, assuming no mutual coupling
C. The value of either inductor, assuming no mutual coupling
D. The total inductance cannot be determined without knowing the exact inductances

The answer is B. The total inductance of inductors in series is equal to the simple sum of all the inductance values, assuming no mutual coupling. See answers 3BE-14.4 and 3BE-14.5.

3BE-14.7 What is the total capacitance of two equal, series-connected capacitors?
A. Half the capacitance of either capacitor
B. Twice the capacitance of either capacitor
C. The value of either capacitor

D. The total capacitance cannot be determined without knowing the exact capacitances

The answer is A. Capacitors in series behave the same as resistors in parallel. See answer 3BE-14.2.

3BE-15.1 What is the voltage across a 500 turn secondary winding in a transformer when the 2250 turn primary is connected to 117-vac?
A. 2369 volts   B. 526.5 volts   C. 26 volts   D. 5.8 volts

The answer is C. The voltage ratio of a transformer is equal to the turns ratio. This is stated mathematically as follows:

$$\frac{T_p}{T_s} = \frac{E_p}{E_s}$$

where: T_p are the turns in the primary winding,
T_s are the turns in the secondary winding,
E_p is the primary voltage,
E_s is the secondary voltage.

In order to find E_s, we must mathematically change the formula so that we can easily work the problem.

$$E_s = \frac{E_p \times T_s}{T_p} \quad \frac{117 \times 500}{2250} = 26 \text{ volts}$$

3BE-15.2 What is the turns ratio of a transformer to match an audio amplifier having an output impedance of 200 ohms to a speaker having an impedance of 10 ohms?
A. 4.47 to 1   B. 14.14 to 1   C. 20 to 1   D. 400 to 1

The answer is A. The turns ratio of a transformer that is used to match the impedance of an amplifier's output to a speaker is found by using the following formula:

$$\text{Turns ratio} = \sqrt{\frac{Z_p}{Z_s}}$$

where: Z_p is the output impedance
Z_s is the speaker impedance

We substitute the values given in the problem to arrive at the answer:

$$\text{Turns ratio} = \sqrt{\frac{200}{10}} = 4.47$$

3BE-15.3 What is the turns ratio of a transformer to match an audio amplifier having an output impedance of 600 ohms to a speaker having an impedance of 4 ohms?
A. 12.2 to 1   B. 24.4 to 1   C. 150 to 1   D. 300 to 1

The answer is A. This problem is solved in the same manner as the above problem.

$$\text{Turns ratio} = \sqrt{\frac{600}{4}} = 12.25$$

3BE-15.4 What is the impedance of a speaker which requires a transformer with a turns ratio of 24 to 1 to match an audio amplifier having an output impedance of 2000 ohms?

ELECTRICAL PRINCIPLES

A. 576 ohms   B. 83.3 ohms   C. 7.0 ohms   D. 3.5 ohms

The answer is D. The basic formula showing the relationship between the turns ratio of an impedance matching transformer and the impedances to be matched, is:

$$TR = \sqrt{\frac{Zp}{Zs}}$$

where: TR is the turns ratio,
Zp is the output impedance,
Zs is the speaker impedance.

Since we are interested in Zs, we must solve for Zs.

$$TR^2 = \frac{Zp}{Zs} \qquad Zs = \frac{Zp}{TR^2}$$

We can now substitute the quantities given in the problem:

$$Zs = \frac{2000}{24^2} = \frac{2000}{576} = 3.47 \text{ ohms.}$$

The impedance of the speaker is, therefore, 3.47 ohms.

3BE-16.1 What is the voltage that would produce the same amount of heat over time in a resistive element as would an applied sine wave ac voltage?
A. A dc voltage equal to the peak-to-peak value of the ac voltage
B. A dc voltage equal to the RMS value of the ac voltage
C. A dc voltage equal to the average value of the ac voltage
D. A dc voltage equal to the peak value of the ac voltage

The answer is B. The DC voltage that will produce the same amount of heat, over time, in a resistive element, as an applied sine wave AC voltage, is known as the ROOT MEAN SQUARE (RMS) voltage. It is also referred to as the EFFECTIVE value of the voltage. The effective value is equal to .707 multiplied by the peak value of the sine wave voltage.

3BE-16.2 What is the peak-to-peak voltage of a sine wave which has an RMS voltage of 117-volts?
A. 82.7 volts   B. 165.5 volts   C. 183.9 volts   D. 330.9 volts

The answer is D. The peak value of a sine wave is equal to 1.414 multiplied by the RMS VOLTAGE. However, the peak value is the value from zero to peak. This is shown in Figure 3BE-16.2. The peak-to-peak value is the value from the very top of the sine wave peak to the bottom of the sine wave peak.

This is equal to 2.828 multiplied by the RMS voltage. The peak-to-peak voltage of this problem is, therefore, equal to:

$$2.828 \times 117 = 330.87 \text{ V.}$$

3BE-16.3 A sine wave of 17-volts peak is equivalent to how many volts RMS?
A. 8.5 volts   B. 12 volts   C. 24 volts   D. 34 volts

The answer is B. The RMS voltage is equal to .707 multiplied by the

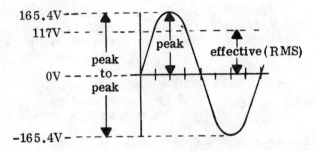

Fig. 3BE-16.2 An AC sine wave.

peak voltage. We substitute the value given in the problem to find the RMS voltage

RMS voltage = 0.707 x 17 = 12.02 volts.

SUBELEMENT 3BF
CIRCUIT COMPONENTS
(1 question)

3BF-1.5 What is the effect of an increase in ambient temperature on the resistance of a carbon resistor?
A. The resistance will increase by 20% for every 10 degrees centigrade that the temperature increases
B. The resistance stays the same
C. The resistance change depends on the resistor's temperature coefficient rating
D. The resistance becomes time dependent

The answer is C. The resistance of carbon decreases with a temperature increase. However, carbon resistors are made of a mixture of carbon and other materials, and the resistance change will depend upon the temperature coefficient of the actual material that the resistor is made of.

3BF-2.6 What type of capacitor is often used in power supply circuits to filter the rectified ac?
A. Disc ceramic   B. Vacuum variable
C. Mica   D. Electrolytic

The answer is D. Most power-supply filters use electrolytic capacitors. Electrolytic capacitors are ideally suited for power supplies because of their small space requirements for a given amount of capacity. Power supply filters require capacitors of high capacitance values. See answer 3AF-2.3.

3BF-2.7 What type of capacitor is used in power supply circuits to filter transient voltage spikes across the transformer secondary winding?
A. High-value   B. Trimmer   C. Vacuum variable   D. Suppressor
The answer is D. Paper capacitors are used.

3BF-3.5 How do inductors become self-resonant?
A. Through distributed electromagnetism
B. Through eddy currents
C. Through distributed capacitance
D. Through parasitic hysteresis

The answer is C. Inductors become self-resonant because there is a certain amount of capacity (distributed capacity) between the turns of wire. When you have a circuit with capacity and inductance, the circuit will be resonant to a certain frequency. The resonant frequency can be found by the following formula:

$$F_R = \frac{1}{2\pi\sqrt{LC}}$$

where: F_R is the resonant frequency
$2\pi = 6.28$
L is the inductance in henries
C is the capacitance in farads.

3BF-4.1 What circuit component can change 120-vac to 400-vac?
A. A transformer   B. A capacitor   C. A diode   D. An SCR

The answer is A. A transformer can change an AC voltage of one value to an AC voltage of another value. Where we wish to change 120 volts AC to 400 volts AC, we would use a step-up transformer with a primary to secondary turns ratio of 3 to 10.

3BF-4.2 What is the source of energy connected to in a transformer?
A. To the secondary winding B. To the primary winding
C. To the core D. To the plates
The answer is B. The primary winding of a transformer is connected to the source of energy. The secondary winding of the transformer is connected to the load.

3BF-4.3 When there is no load attached to the secondary winding of a transformer, what is current in the primary winding called?
A. Magnetizing current B. Direct current
C. Excitation current D. Stabilizing current
The answer is A.

3BF-4.4 In what terms are the primary and secondary windings ratings of a power transformer usually specified?
A. Joules per second B. Peak inverse voltage
C. Coulombs per second D. Volts or volt-amperes
The answer is D. They are usually specified in volts and amperes, or volts and milliamperes. There are other transformer winding specifications that include ohms and turns per square inch.

3BF-5.1 What is the peak-inverse-voltage rating of a power supply rectifier?
A. The highest transient voltage the diode will handle
B. 1.4 times the AC frequency
C. The maximum voltage to be applied in the non-conducting direction
D. 2.8 times the AC frequency
The answer is C. The peak-inverse-voltage (PIV) rating is the maximum voltage that can be placed across the diode in its reverse polarity state (anode is negative with respect to cathode). If the PIV is exceeded, there may be a breakdown which may destroy the diode.

3BF-5.2 Why must silicon rectifier diodes be thermally protected?
A. Because of their proximity to the power transformer
B. Because they will be destroyed if they become too hot
C. Because of their susceptibility to transient voltages
D. Because of their use in high-voltage applications
The answer is B. They must be thermally protected because they are physically small and carry high currents (high current density). Some silicon diodes are built with efficient heat sinks to carry the heat away from the junction.

3BF-5.4 What are the two major ratings for silicon diode rectifiers of the type used in power supply circuits which must not be exceeded?
A. Peak load impedance; peak voltage
B. Average power; average voltage
C. Capacitive reactance; avalanche voltage
D. Peak inverse voltage; average forward current
The answer is D. The peak inverse voltage is discussed in 3BF-5.1. The average forward current is the average current drawn by the load.

SUBELEMENT 3BG
PRACTICAL CIRCUITS
(1 question)

3BG-1.1 Why should a resistor and capacitor be wired in parallel with power-supply rectifier diodes?
A. To equalize voltage drops and guard against transient voltage spikes
B. To ensure that the current through each diode is about the same
C. To smooth the output waveform
D. To decrease the output voltage

The answer is A. Rectifier diodes can be wired in series in order to obtain a high peak-inverse-voltage rating. When this is done, a resistor is placed across each diode to equalize the voltage drops across the diodes. A capacitor may also be placed across each diode to protect it against high voltage transients and "spikes".

3BG-1.2 What function do capacitors serve when resistors and capacitors are connected in parallel with high voltage power supply rectifier diodes?
A. They double or triple the output voltage
B. They block the alternating current
C. They protect those diodes that develop back resistance faster than other diodes
D. They regulate the output voltage

The answer is C. See answer 3BG-1.1.

3BG-1.3 What is the output waveform of an unfiltered full wave rectifier connected to a resistive load?
A. A steady dc voltage
B. A sine wave at half the frequency of the ac input
C. A series of pulses at the same frequency as the ac input
D. A series of pulses at twice the frequency of the ac input

The answer is D. Figure 3BG-1.3 illustrates a full wave rectifier, together with the AC waveform across the transformer secondary, and the output waveform across the load. The output waveform is a pulsating DC wave. The entire wave is above the 0 line; therefore, it is DC.

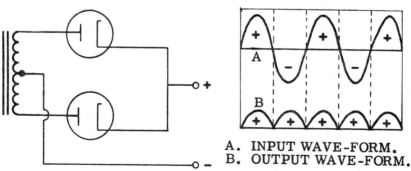

A. INPUT WAVE-FORM.
B. OUTPUT WAVE-FORM.

Figure 3BG-1.3. Full-wave rectifier with wave forms.

3BG-1.4 How many degrees of each cycle does a half-wave rectifier utilize?

A. 90 degrees   B. 180 degrees   C. 270 degrees   D. 360 degrees

The answer is B. Figure 3BG-1.4 illustrates a half-wave rectifier, together with the AC waveform across the transformer secondary, and the output waveform across the load. Note that one half of the input wave is reproduced. Therefore, half of 360 degrees or 180 degrees is utilized.

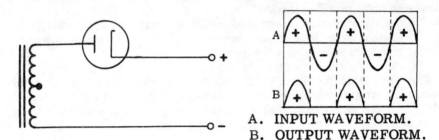

A. INPUT WAVEFORM.
B. OUTPUT WAVEFORM.

Figure 3BG-1.4. Half-wave rectifier with wave forms.

3BG-1.5 How many degrees of each cycle does a full-wave rectifier utilize?
A. 90 degrees   B. 180 degrees   C. 270 degrees   D. 360 degrees

The answer is D. We can see from Figure 3BG-1.3 that output current flows during the entire cycle. The negative halves of the input waveform are merely inverted. Therefore, 360 degrees of each cycle are utilized in a full-wave rectifier system.

3BG-1.6 Where is a power supply bleeder resistor connected?
A. Across the filter capacitor
B. Across the power-supply input
C. Between the transformer primary and secondary
D. Across the inductor in the output filter

The answer is A. A bleeder resistor is placed across the output of the power supply. See Figure 3BG-1.7.

3BG-1.7 What components comprise a power supply filter network?
A. Diodes   B. Transformers and transistors
C. Quartz crystals   D. Capacitors and inductors

The answer is D. A power supply filter network usually consists of one or more capacitors and one or more inductors. Figure 3BG-1.7 shows how a practical filter is connected in a power supply.

3BG-1.8 What should be the peak-inverse-voltage rating of the rectifier in a full-wave power supply?
A. One-quarter the normal output voltage of the power supply
B. Half the normal output voltage of the power supply
C. Equal to the normal output voltage of the power supply
D. Double the normal peak output voltage of the power supply

The answer is D. See answer 3BF-5.1. In the case of a full wave rectifier, the peak-inverse-voltage is the peak voltage of the entire secondary winding of the transformer. The peak-inverse-voltage is twice the normal peak output voltage of the power supply because the output voltage is derived from one half of the entire secondary winding.

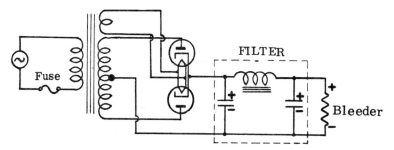

Figure 3BG-1.7. Full-wave rectifier with filter.

3BG-1.9 What should be the peak-inverse-voltage rating of the rectifier in a half-wave power supply?
A. One-quarter to one-half the normal peak output voltage of the power supply
B. Half the normal output voltage of the power supply
C. Equal to the normal output voltage of the power supply
D. One to two times the normal peak output voltage of the power supply

The answer is D. See answer 3BG-1.8. If the power supply has a resistive or inductive load, the peak-inverse-voltage is equal to the peak voltage of the entire secondary winding of the transformer which is approximately equal to the normal peak output voltage of the power supply. If the output of the power supply has a capacitive load, the peak-inverse-voltage should be approximately twice the peak output voltage of the power supply or twice the peak voltage of the secondary winding of the transformer. This is because the filter capacitor charges up to the peak value of the secondary voltage and is in series with the secondary voltage and the rectifier.

3BG-2.8 What should the impedance of a low-pass filter be as compared to the impedance of the transmission line into which it is inserted?
A. Substantially higher
B. About the same
C. Substantially lower
D. Twice the transmission line impedance

The answer is B. For proper operation, the impedance of the low-pass filter should equal the impedance of the transmission line into which it is inserted.

SUBELEMENT 3BH
SIGNALS AND EMISSIONS
(2 questions)

3BH-2.1 What is the term for alteration of the amplitude of an rf wave for the purpose of conveying information?
A. Frequency modulation
B. Phase modulation
C. Amplitude rectification
D. Amplitude modulation

The answer is D. In amplitude modulation, the audio signal to be transmitted is superimposed onto the RF carrier by varying the amplitude of the RF carrier in accordance with the audio signal.

3BH-2.3 What is the term for alteration of the phase of an rf wave for the purpose of conveying information?
A. Pulse modulation
B. Phase modulation
C. Phase rectification
D. Amplitude modulation

The answer is B. Phase modulation is slightly different from frequency modulation. In frequency modulation we directly alter the frequency of the RF carrier in accordance with the audio. In phase modulation we alter the phase of the carrier which in turn alters the frequency of the carrier. Phase modulation is sometimes called indirect FM. Actually, both phase and frequency modulation are types of FM.

3BH-2.4 What is the term for alteration of the frequency of an rf wave for the purpose of conveying information?
A. Phase rectification
B. Frequency rectification
C. Amplitude modulation
D. Frequency modulation

The answer is D.

3BH-3.1 In what emission type does the instantaneous amplitude (envelope) of the rf signal vary in accordance with the modulating af?
A. Frequency shift keying
B. Pulse modulation
C. Frequency modulation
D. Amplitude modulation

The answer is D. See answer 3BH-2.1.

3BH-3.2 What determines the spectrum space occupied by each group of sidebands generated by a correctly operating emission A3E transmitter?
A. The audio frequencies used to modulate the transmitter
B. The phase angle between the audio and radio frequencies being mixed
C. The radio frequencies used in the transmitter's VFO
D. The CW keying speed

The answer is A. The bandwidth or spectrum space of a correctly operating AM transmitter is determined by the frequency of the audio signal. The higher the frequency, the wider the spectrum space occupied by the sideband frequencies.

3BH-4.1 How much is the carrier suppressed in an emission J3E transmission?
A. No more than 20 dB below peak output power
B. No more than 30 dB below peak output power
C. At least 40 dB below peak output power
D. At least 60 dB below peak output power

The answer is C.

3BH-4.2 What is one advantage of carrier suppression in an emission A3E transmission?
A. Only half the bandwidth is required for the same information content
B. Greater modulation percentage is obtainable with lower distortion
C. More power can be put into the sidebands
D. Simpler equipment can be used to receive a double-sideband suppressed carrier signal

The answer is C. A double-sideband suppressed-carrier signal requires only one third of the power for similar results, compared to full carrier AM. This is because two thirds of the power is in the carrier, which is suppressed, and one third is in the sidebands.

3BH-5.1 Which one of the telephony emissions popular with amateurs occupies the narrowest band of frequencies?
A. Single-sideband emission   B. Double-sideband emission
C. Phase-modulated emission   D. Frequency-modulated emission

The answer is A. Single sideband emission normally has the narrowest bandwidth. Half of the bandwidth of double sideband is suppressed.

3BH-5.2 Which emission type is produced by a telephony transmitter having a balanced modulator followed by a 2.5-kHz bandpass filter?
A. PM   B. AM   C. SSB   D. FM

The answer is C. Single sideband, suppressed carrier emission is produced. Figure 3BH-5.2 shows a block diagram of an SSB transmitter.

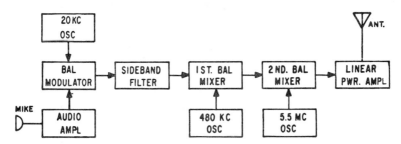

Figure 3BH-5.2. Block diagram of an SSB transmitter.

3BH-7.2 What emission is produced by a reactance modulator connected to an rf power amplifier?
A. Multiplex modulation   B. Phase modulation
C. Amplitude modulation   D. Pulse modulation

The answer is B. When the reactance modulator is connected to an AMPLIFIER the output will be phase modulated. However, when the reactance modulator is connected to an OSCILLATOR, the output will be frequency modulated. Actually, frequency modulation and phase modulation are similar to each other, since one does not exist without the other.

3BH-8.1 What purpose does the carrier serve in an emission A3E transmission?
A. The carrier separates the sidebands so they don't cancel in the receiver
B. The carrier contains the modulation information

C. The carrier maintains symmetry of the sidebands to prevent distortion
D. The carrier serves as a reference signal for demodulation by an envelope detector

The answer is D. It serves as a reference frequency for the modulated emissions. Once the signal reaches the receiver detector circuit, the audio intelligence is extracted from the signal, using the carrier.

3BH-8.2 What signal component appears in the center of the frequency band of an emission A3E transmission?
A. The lower sidebands B. The subcarrier
C. The carrier D. The pilot tone

The answer is C. The carrier is at the center of the emitted bandwidth and the sidebands appear on both sides of the carrier.

3BH-9.1 What sidebands are generated by an emission A3E transmitter with a 7250-kHz carrier modulated less than 100% by an 800-Hz pure sine wave?
A. 7250.8 kHz and 7251.6 kHz B. 7250.0 kHz and 7250.8 kHz
C. 7249.2 kHz and 7250.8 kHz D. 7248.4 kHz and 7249.2 kHz

The answer is C. The two new sideband frequencies will be:
(1) 7,250 kHz + 800 Hz (0.8 kHz) = 7,250.8 kHz
(2) 7,250 kHz - 800 Hz (0.8 kHz) = 7,249.2 kHz

3BH-10.1 How many times over the maximum deviation is the bandwidth of an emission F3E transmission?
A. 1.5 B. at least 2.0 C. at least 4.0
D. The bandwidth cannot be determined without knowing the exact carrier and modulating frequencies involved

The answer is B. In AM, a single audio frequency modulating signal causes a single sideband on each side of the carrier. In FM, a single audio frequency modulating signal causes many sidebands on either side of the center frequency. FM, therefore, occupies a greater bandwidth than AM or than the frequency deviation would indicate.

3BH-10.2 What is the total bandwidth of an emission F3E transmission having 5-kHz deviation and 3-kHz af?
A. 3 kHz B. 5 kHz C. 8 kHz D. 16 kHz

The answer is D. The total bandwidth of an FM signal is equal to twice the sum of the deviation and the maximum audio modulating frequency.
Bandwidth = 2 x (5 + 3) = 16 kHz.

3BH-11.1 What happens to the shape of the rf envelope, as viewed on an oscilloscope, of an emission A3E transmission?
A. The amplitude of the envelope increases and decreases in proportion to the modulating signal
B. The amplitude of the envelope remains constant
C. The brightness of the envelope increases and decreases in proportion to the modulating signal
D. The frequency of the envelope increases and decreases in proportion to the amplitude of the modulating signal

The answer is A. The envelope's shape varies in accordance with the frequency and amplitude of the modulating signal.

SIGNALS AND EMISSIONS

3BH-13.1 What results when an emission J3E transmitter is overmodulated?
A. The signal becomes louder with no other effects
B. The signal occupies less bandwidth with poor high frequency response
C. The signal has higher fidelity and improved signal-to-noise ratio
D. The signal becomes distorted and occupies more bandwidth

The answer is D. The result of overmodulation is excessive bandwidth and distortion. It causes interference or "splatter" on nearby frequencies.

3BH-13.2 What results when an emission A3E transmitter is overmodulated?
A. The signal becomes louder with no other effects
B. The signal becomes distorted and occupies more bandwidth
C. The signal occupies less bandwidth with poor high frequency response
D. The transmitter's carrier frequency deviates

The answer is B. See answer 3BH-13.1.

3BH-15.1 What is the frequency deviation for a 12.21-MHz reactance-modulated oscillator in a 5-kHz deviation, 146.52-MHz F3E transmitter?
A. 41.67 Hz   B. 416.7 Hz   C. 5 kHz   D. 12 kHz

The answer is B. First we divide the output frequency of the FM transmitter by the oscillator frequency to determine the amount of frequency multiplication of the transmitter.

146.52 MHz divided by 12.21 MHz = 12

Then we divide the output frequency deviation by 12 to get the deviation at the oscillator output:

5 kHz (5000 Hz) divided by 12 = 416.66 Hz.

3BH-15.2 What stage in a transmitter would translate a 5.3-MHz input signal to 14.3-MHz?
A. A mixer
B. A beat frequency oscillator
C. A frequency multiplier
D. A linear translator stage

The answer is A. The mixer stage would have received the 5.3 MHz signal and a local oscillator signal of 9 MHz or 19.6 MHz to give us the sum or difference of 14.3 MHz. It could not be a frequency multiplier stage because the output would have been an exact multiple of the input, such as 10.6 MHz or 15.9 MHz.

3BH-16.4 How many frequency components are in the signal from an af shift keyer at any instant?
A. One   B. Two   C. Three   D. Four

The answer is A.

3BH-16.5 How is frequency shift related to keying speed in an fsk signal?
A. The frequency shift in Hertz must be at least four times the keying speed in WPM
B. The frequency shift must not exceed 15 Hz per WPM of keying speed
C. Greater keying speeds require greater frequency shifts
D. Greater keying speeds require smaller frequency shifts

The answer is C.

SUBELEMENT 3BI
ANTENNAS AND FEEDLINES
(4 questions)

3BI-1.3 Why is a Yagi antenna often used for radiocommunications on the 20 meter band?
A. It provides excellent omnidirectional coverage in the horizontal plane
B. It is smaller, less expensive and easier to erect than a dipole or vertical antenna
C. It discriminates against interference from other stations off to the side or behind
D. It provides the highest possible angle of radiation for the HF bands

The answer is C. The 14 MHz band is used for long distance communications, and it is important to have high transmitting and receiving gain in the direction of the station being worked.

3BI-1.7 What method is best suited to match an unbalanced coaxial feed line to a Yagi antenna?
A. "T" match   B. Delta match   C. Hairpin match   D. Gamma match

The answer is D. A GAMMA MATCH is used to connect an unbalanced coaxial feedline to the balanced driven element of a Yagi antenna. A Gamma Match consists of a length of coaxial feedline, a rod with a movable clamp, and a variable resonating capacitor. The capacitor tunes out the reactance of the Gamma rod. Figure 3BI-1.7 illustrates a Gamma Match.

Fig. 3BI-1.7. A Gamma Match.

3BI-1.9 How can the bandwidth of a parasitic beam antenna be increased?
A. Use larger diameter elements   B. Use closer element spacing
C. Use traps on the elements   D. Use tapered-diameter elements

The answer is A. The bandwidth of an antenna depends upon the diameter of the antenna elements and the radiation resistance of the antenna. The larger the diameter of the elements, the greater is the bandwidth. The bandwidth is also increased if the radiation resistance is increased. We can increase the radiation resistance by increasing the spacing between the elements of the antenna.

3BI-2.1 How much gain over a half-wave dipole can a two-element cubical quad antenna provide?
A. Approximately 0.6 dB   B. Approximately 2 dB
C. Approximately 6 dB   D. Approximately 12 dB
The answer is C.

ANTENNAS AND FEEDLINES

3BI-3.1 How long is each side of a cubical quad antenna driven element for 21.4-MHz?
A. 1.17 feet   B. 11.7 feet   C. 47 feet   D. 469 feet

The answer is B. We use the following formula for determining the total length of the driven element of a Quad antenna:

$$\text{Length (feet)} = \frac{1005}{\text{frequency in MHz.}} = \frac{1005}{21.4} = 46.96 \text{ feet}$$

We then divide 46.96 feet by 4 and we obtain 11.74 feet, which is the length of each side of the driven element of the Quad antenna.

3BI-3.2 How long is each side of a cubical quad antenna driven element for 14.3-MHz?
A. 1.75 feet   B. 17.6 feet   C. 23.4 feet   D. 70.3 feet

The answer is B. We use the same formula as in answer 3BI-3.1 for determining the total length of the driven element of a Quad antenna:

$$\text{Length (feet)} = \frac{1005}{14.3} = 70.28 \text{ feet}$$

We then divide 70.28 feet by 4 and we obtain 17.57 feet, which is the length of each side of the driven element of the quad antenna.

3BI-3.3 How long is each side of a cubical quad antenna reflector element for 29.6-MHz?
A. 8.23 feet   B. 8.7 feet   C. 9.7 feet   D. 34.8 feet

The answer is B. The formula for finding the total length of the reflector element of a cubical quad antenna is:

$$\text{Length (feet)} = \frac{1030}{\text{frequency in MHz.}} = \frac{1030}{29.6} = 34.8 \text{ feet}$$

We then divide 34.8 feet by 4 and obtain 8.7 feet, which is the length of each side of the reflector element of the quad antenna.

3BI-3.4 How long is each leg of a symmetrical delta loop antenna driven element for 28.7-MHz?
A. 8.75 feet   B. 11.32 feet   C. 11.7 feet   D. 35 feet

The answer is C. The delta loop antenna is similar to the quad antenna in that the total length of the elements of both of them are a full wavelength long. However, the delta loop antenna has three sides whereas the quad antenna has four sides. The total length of the driven element of a delta loop antenna is:

$$\text{Length (feet)} = \frac{1005}{\text{frequency in MHz.}} = \frac{1005}{28.7} = 35.02 \text{ feet}$$

We then divide 35 feet by 3 to obtain 11.67 feet, which is the length of each leg of the driven element of the delta loop antenna.

3BI-3.5 How long is each leg of a symmetrical delta loop antenna driven element for 24.9-MHz?
A. 10.09 feet   B. 13.05 feet   C. 13.45 feet   D. 40.36 feet

The answer is C. The formula used for finding the total length of the driven element of a delta loop antenna is:

$$\text{Length (feet)} = \frac{1005}{\text{frequency in MHz}} = \frac{1005}{24.9} = 40.36 \text{ feet}$$

We then divide 40.36 feet by 3 to obtain 13.45 feet, which is the length of each side of the diven element of the delta loop antenna.

3BI-3.6 How long is each leg of a symmetrical delta loop antenna reflector element for 14.1-MHz?
A. 18.26 feet B. 23.76 feet
C. 24.35 feet D. 73.05 feet

The answer is C. The formula for finding the approximate total length of the reflector element is:

$$\text{Length (feet)} = \frac{1030}{\text{frequency in MHz}} = \frac{1030}{14.1} = 73.05 \text{ feet}$$

We then divide 73.05 feet by 3 to obtain 24.35 feet, which is the length of each side of the reflector element of a delta loop antenna.

3BI-3.7 How long is the driven element of a Yagi antenna for 14.0-MHz?
A. Approximately 17 feet B. Approximately 33 feet
C. Approximately 35 feet D. Approximately 66 feet

The answer is B. The formula for the approximate length in feet of the driven element of a Yagi antenna is:

$$\text{Length (feet)} = \frac{472}{\text{frequency in MHz}} = \frac{472}{14.0} = 33.71 \text{ feet}$$

3BI-3.8 How long is the director element of a Yagi antenna for 21.1-MHz?
A. Approximately 42 feet B. Approximately 21 feet
C. Approximately 17 feet D. Approximately 10.5 feet

The answer is B. The formula for the approximate length in feet of the director element of a Yagi antenna is:

$$\text{Length (feet)} = \frac{458}{\text{frequency in MHz}} = \frac{458}{21.1} = 21.7 \text{ feet}$$

3BI-3.9 How long is the reflector element of a Yagi antenna for 28.1MHz?
A. Approximately 8.75 feet B. Approximately 16.6 feet
C. Approximately 17.5 feet D. Approximately 35 feet

The answer is C. The formula for the approximate length in feet of the reflector element of a Yagi antenna is:

$$\text{Length (feet)} = \frac{490}{\text{frequency in MHz}} = \frac{490}{28.1} = 17.44 \text{ feet}$$

3BI-5.1 What is the feedpoint impedance for a half-wavelength dipole HF antenna suspended horizontally one-quarter wavelength or more above the ground?
A. Approximately 50 ohms, resistive
B. Approximately 73 ohms, resistive and inductive
C. Approximately 50 ohms, resistive and capacitive
D. Approximately 73 ohms, resistive

ANTENNAS AND FEEDLINES

The answer is D. It is 73 ohms at 1/4 wavelength above ground. It will change if the distance above ground is changed.

3BI-5.2 What is the feedpoint impedance of a quarter-wavelength vertical HF antenna with a horizontal ground plane?
A. Approximately 18 ohms
B. Approximately 36 ohms
C. Approximately 52 ohms
D. Approximately 72 ohms
The answer is B.

3BI-5.3 What is an advantage of downward sloping radials on a ground-plane antenna?
A. Sloping the radials downward lowers the radiation angle
B. Sloping the radials downward brings the feed point impedance close to 300 ohms
C. Sloping the radials downward allows rainwater to run off the antenna
D. Sloping the radials downward brings the feed point impedance closer to 50 ohms.

The answer is D. Downward sloping radials increase the feedpoint impedance closer to 50 ohms. This makes it easier for the antenna to match the impedance of most transmission lines. When the radials are not sloping the impedance is approximately 30 to 36 ohms.

3BI-5.4 What happens to the feedpoint impedance of a ground-plane antenna when the radials slope downward from the base of the antenna?
A. The feed point impedance decreases
B. The feed point impedance increases
C. The feed point impedance stays the same
D. The feed point impedance becomes purely capacitive
The answer is B. See answer 3BI-5.3.

3BI-6.1 Compared to a dipole antenna, what are the directional radiation characteristics of a cubical quad HF antenna?
A. The quad has more directivity in the horizontal plane but less directivity in the vertical plane
B. The quad has less directivity in the horizontal plane but more directivity in the vertical plane
C. The quad has more directivity in both horizontal and vertical planes
D. The quad has less directivity in both horizontal and vertical planes
The answer is C. The Quad antenna has more gain and directivity in both the horizontal and vertical directions than a comparable dipole.

3BI-6.2 What is the radiation pattern of an ideal half-wavelength dipole HF antenna?
A. If it is installed parallel to the earth, it radiates well in a figure-eight pattern at right angles to the antenna wire
B. If it is installed parallel to the earth, it radiates well in a figure-eight pattern off both ends of the antenna wire
C. If it is installed parallel to the earth, it radiates equally well in all directions
D. If it is installed parallel to the earth, the pattern will have two lobes on one side of the antenna wire, and one larger lobe on the other side

The answer is A. An ideal half-wave dipole has a figure 8 radiation pattern. This is shown in Figure 3BI-6.2. It radiates well in the two directions perpendicular to the length of the antenna. It radiates poorly in the two directions along the length of the antenna.

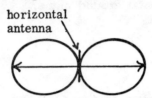

Figure 3BI-6.2. Radiation pattern of a horizontal half-wave antenna as viewed from above.

3BI-6.3 How does proximity to the ground affect the radiation pattern of a horizontal dipole HF antenna?
A. If the antenna is too far from the ground, the pattern becomes unpredictable
B. If the antenna is less than one-half wavelength from the ground, reflected radio waves from the ground distort the radiation pattern of the antenna
C. A dipole antenna's radiation pattern is unaffected by its distance to the ground
D. If the antenna is less than one-half wavelength from the ground, radiation off the ends of the wire is reduced

The answer is B. In order for the radiation pattern of an antenna to be close to the ideal pattern, the antenna must be located at least one-half wavelength above ground. If it is less than one-half wavelength above ground, the directional pattern of the antenna will be distorted.

3BI-6.4 What does the term antenna front-to-back ratio mean?
A. The number of directors versus the number of reflectors
B. The relative position of the driven element with respect to the reflectors and directors
C. The power radiated in the major radiation lobe compared to the power radiated in exactly the opposite direction
D. The power radiated in the major radiation lobe compared to the power radiated 90 degrees away from that direction

The answer is C. The "front-to-back" ratio of an antenna is the amount of power radiated in the direction of maximum radiation, to the amount of power radiated in the exact opposite (180 degree) direction.

3BI-6.5 What effect upon the radiation pattern of a HF dipole antenna, will a slightly smaller parasitic parallel element located a few feet away in the same horizontal plane have?
A. The radiation pattern will not change appreciably
B. A major lobe will develop in the horizontal plane, parallel to the two elements
C. A major lobe will develop in the vertical plane, away from the ground
D. If the spacing is greater than 0.1 wavelength, a major lobe will develop in the horizontal plane to the side of the driven element toward the parasitic element

ANTENNAS AND FEEDLINES

The answer is D. The addition of a director element causes the radiation to be concentrated more in one direction. This changes the radiation pattern from the figure 8 type shown in Figure 3BI-6.2, to the one shown in Figure 3BI-6.6.

3BI-6.6 What is the meaning of the term main lobe as used in reference to a directional antenna?
A. The direction of least radiation from an antenna
B. The point of maximum current in a radiating antenna element
C. The direction of maximum radiated field strength from a radiating antenna
D. The maximum voltage standing wave point on a radiating element

The answer is C. The RF energy radiated from a beam antenna is greater in one direction than it is in the other directions. To depict this visually, a graph, called the radiation pattern, is used. See Figure 3BI-6.6.

On the diagram, each egg-shaped protrusion is called a lobe. The largest is the MAJOR or MAIN lobe; all others are referred to as MINOR lobes. The antenna is represented as a small dot in the center of the lobes.

The diagram shows the horizontal pattern, a view looking down on the antenna from above the earth. If, as shown, lines are drawn outward from the antenna to the edges of the lobes, their lengths are proportional to the signal strengths in those directions. In this example, maximum energy in the major lobe is to the North. The arrow pointing northeast shows considerably less power in that direction.

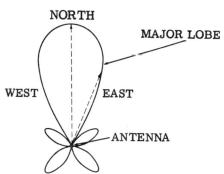

Figure 3BI-6.6. Horizontal pattern of a Yagi antenna, as viewed from top.

3BI-7.1 Upon what does the characteristic impedance of a parallel conductor antenna feedline depend?
A. The distance between the centers of the conductors and the radius of the conductors
B. The distance between the centers of the conductors and the length of the line
C. The radius of the conductors and the frequency of the signal
D. The frequency of the signal and the length of the line

The answer is A. The characteristic impedance of an air-insulated parallel-conductor transmission line is dependent upon the radius of the conductors and the distance between the conductors. Its formula is:

$$z = 276 \log \frac{b}{a}$$

Where: b is the distance between conductors
a is the radius of the conductors.

To a lesser degree, the type of dielectric used to separate the conductors also affects the characteristic impedance.

3BI-7.2 What is the characteristic impedance of various coaxial cables commonly used for antenna feedlines at amateur stations?
A. Around 25 and 30 ohms
B. Around 50 and 75 ohms
C. Around 80 and 100 ohms
D. Around 500 and 750 ohms
The answer is B.

3BI-7.3 What effect, if any, does the length of a coaxial cable have upon its characteristic impedance?
A. The length has no effect on the characteristic impedance
B. The length affects the characteristic impedance primarily above 144MHz
C. The length affects the characteristic impedance primarily below 144MHz
D. The length affects the characteristic impedance at any frequency
The answer is A.

3BI-7.4 What is the characteristic impedance of flat-ribbon TV-type twinlead?
A. 50 ohms   B. 75 ohms   C. 100 ohms   D. 300 ohms
The answer is D.

3BI-8.4 What is the cause of power being reflected back down an antenna feedline?
A. Operating an antenna at its resonant frequency
B. Using more transmitter power than the antenna can handle
C. A difference between feed line impedance and antenna feed point impedance
D. Feeding the antenna with unbalanced feed line
The answer is C. See answer 3AI-8.3.

3BI-9.3 What will be the standing wave ratio when a 50 ohm feed line is connected to a resonant antenna having a 200 ohm feedpoint impedance?
A. 4:1   B. 1:4   C. 2:1   D. 1:2
The answer is A. The standing wave ratio is equal to the larger impedance, divided by the smaller impedance:

$$\text{SWR} = \frac{200}{50} = \frac{4}{1} \quad \text{or} \quad 4 \text{ to } 1$$

3BI-9.4 What will be the standing wave ratio when a 50 ohm feed line is connected to a resonant antenna having a 10 ohm feedpoint impedance?
A. 2:1   B. 50:1   C. 1:5   D. 5:1
The answer is D. The standing wave ratio is equal to the larger impedance, divided by the smaller impedance.

3BI-9.5 What will be the standing wave ratio when a 50 ohm feed line is connected to a resonant antenna having a 50 ohm feedpoint impedance?
A. 2:1   B. 50:50   C. 1:1   D. 0:0
The answer is C. The standing wave ratio is equal to the larger impedance, divided by the smaller impedance. The SWR is 1:1 and is achieved when the characteristic impedance of the line is equal to the resistance at the antenna feedpoint. This is an ideal situation.

3BI-11.1 How does the characteristic impedance of a coaxial cable affect

the amount of attenuation to the rf signal passing through it?
A. The attenuation is affected more by the characteristic impedance at frequencies above 144 MHz than at frequencies below 144 MHz
B. The attenuation is affected less by the characteristic impedance at frequencies above 144 MHz than at frequencies below 144 MHz
C. The attenuation related to the characteristic impedance is about the same at all amateur frequencies below 1.5 GHz
D. The difference in attenuation depends on the emission type in use

The answer is C. The amount of attenuation is not dependent upon the characteristic impedance of the cable. It is dependent upon the quality and type of dielectric, the thickness of the conductors, the length of line, and similar factors.

3BI-11.2 How does the amount of attenuation to a 2 meter signal passing through a coaxial cable differ from that to a 160 meter signal?
A. The attenuation is greater at 2 meters
B. The attenuation is less at 2 meters
C. The attenuation is the same at both frequencies
D. The difference in attenuation depends on the emission type in use

The answer is A. As the frequency increases, the attenuation increases. This is why loss specifications of transmission lines give the frequency at which the losses are measured.

3BI-11.4 What is the effect on its attenuation when flat-ribbon TV-type twinlead is wet?
A. Attenuation decreases slightly   B. Attenuation remains the same
C. Attenuation decreases sharply   D. Attenuation increases

The answer is D.

3BI-11.7 Why might silicone grease or automotive car wax be applied to flat-ribbon TV-type twinlead?
A. To reduce "skin effect" losses on the conductors
B. To reduce the build up of dirt and moisture on the feedline
C. To increase the velocity factor of the feedline
D. To help dissipate heat during high-SWR operation

The answer is B.

3BI-11.8 In what values are rf feed line losses usually expressed?
A. Bels/1000 ft.   B. DB/1000 ft.   C. Bels/100 ft.   D. DB/100 ft.

The answer is D. It is usually expressed in decibels (dB) per unit length, usually dB per 100 feet.

3BI-11.10 As the operating frequency increases, what happens to the dielectric losses in a feed line?
A. The losses decrease   B. The losses decrease to zero
C. The losses remain the same   D. The losses increase

The answer is D.

3BI-11.12 As the operating frequency decreases, what happens to the dielectric losses in a feed line?
A. The losses decrease   B. The losses increase
C. The losses remain the same   D. The losses become infinite

The answer is A.

3BI-12.1 What condition must be satisfied to prevent standing waves of voltage and current on an antenna feedline?
A. The antenna feedpoint must be at dc ground potential
B. The feedline must be an odd number of electrical quarter wavelengths long
C. The feedline must be an even number of physical half wavelengths long
D. The antenna feedpoint impedance must be matched to the characteristic impedance of the feedline

The answer is D. The impedance of the transmission line must be equal to the impedance of the antenna at the point where the feedline connects to the antenna.

3BI-12.2 How is an inductively-coupled matching network used in an antenna system consisting of a center-fed resonant dipole and coaxial feed line?
A. An inductively coupled matching network is not normally used in a resonant antenna system
B. An inductively coupled matching network is used to increase the SWR to an acceptable level
C. An inductively coupled matching network can be used to match the unbalanced condition at the transmitter output to the balanced condition required by the coaxial line
D. An inductively coupled matching network can be used at the antenna feed point to tune out the radiation resistance

The answer is A. The inductively coupled matching network is called a balun (BALanced to UNbalanced). It matches the balanced antenna to the unbalanced coaxial transmission line. The balun is placed at the point where the transmission line feeds the antenna. The balun also prevents the coaxial cable from improperly radiating RF energy.

3BI-12.5 What is an antenna-transmission line mismatch?
A. A condition where the feed point impedance of the antenna does not equal the output impedance of the transmitter
B. A condition where the output impedance of the transmitter does not equal the characteristic impedance of the feed line
C. A condition where a half-wavelength antenna is being fed with a transmission line of some length other than one-quarter wavelength at the operating frequency
D. A condition where the characteristic impedance of the feed line does not equal the feed point impedance of the antenna

The answer is D. An antenna transmission-line mismatch increases the SWR which reduces the efficiency of the antenna system.

APPENDIX 1
RST Reporting System

The RST Reporting System is a means of rating the quality of a signal on a numerical basis. In this system, the R stands for readabilty, and is rated on a scale of 1 to 5. The S stands for signal strength, and it is rated on a scale of 1 to 9. The T indicates the quality of a CW tone, and its scale is also 1 to 9. The higher the number, the better the signal.

READABILITY

1. Unreadable
2. Barely readable; occasional words distinguishable
3. Readable with considerable difficulty
4. Readable with practically no difficulty
5. Perfectly readable

SIGNAL STRENGTH

1. Faint; signals barely perceptible
2. Very weak signals
3. Weak signals
4. Fair signals
5. Fairly good signals
6. Good signals
7. Moderately strong signals
8. Strong signals
9. Extremely strong signals

TONE

1. Extremely rough, hissing tone
2. Very rough AC note; no trace of musicality
3. Rough, low-pitched AC note; slightly musical
4. Rather rough AC note; moderately musical
5. Musically modulated
6. Modulated note; slight trace of whistle
7. Near DC note; smooth ripple
8. Good DC note; just a trace of ripple
9. Purest DC note (If note appears to be crystal controlled, add letter X after the number indicating tone)

EXAMPLE: Your signals are RST 599X. (Your signals are perfectly readable, extremely strong, have purest DC note, and sound as if your transmitter is crystal-controlled.)

FREQUENCY ALLOCATIONS FOR POPULAR AMATEUR BANDS
All in MegaHertz.   "X" indicates no privileges.

CLASSES	NOVICE		TECHNICIAN		GENERAL AND CONDITIONAL		ADVANCED		EXTRA	
BANDS	CW	PHONE	CW	PHONE	CW	PHONE	CW	PHONE	CW	PHONE
80 METERS	3.7 to 3.75	X	3.7 to 3.75	X	3.525 to 3.750 and 3.85 to 4.0	3.85 to 4.0	3.525 to 3.750 and 3.775 to 4.0	3.775 to 4.0	3.5 to 4.0	3.75 to 4.0
40 METERS	7.1 to 7.15	X	7.1 to 7.15	X	7.025 to 7.150 and 7.225 to 7.3	7.225 to 7.3	7.025 to 7.3	7.15 to 7.3	7.0 to 7.3	7.15 to 7.3
20 METERS	X	X	X	X	14.025 to 14.15 and 14.225 to 14.35	14.225 to 14.35	14.025 to 14.15 and 14.175 to 14.35	14.175 to 14.35	14.0 to 14.35	14.15 to 14.35
15 METERS	21.1 to 21.2	X	21.1 to 21.2	X	21.025 to 21.20 and 21.30 to 21.450	21.3 to 21.450	21.025 to 21.20 and 21.225 to 21.45	21.225 to 21.450	21.0 to 21.450	21.2 to 21.45
10 METERS	28.1 to 28.5	28.3 to 28.5	28.1 to 28.5	28.3 to 28.5	28.0 to 29.7	28.3 to 29.7	28.0 to 29.7	28.3 to 29.7	28.0 to 29.7	28.3 to 29.7
6 METERS	X	X	50.0 to 54.0	50.1 to 54.0	50.0 to 54.0	50.1 to 54.0	50.0 to 54.0	50.1 to 54.0	50.0 to 54.0	50.1 to 54.0
2 METERS	X	X	144.0 to 148.0	144.1 to 148.0	144.0 to 148.0	144.1 to 148.0	144.0 to 148.0	144.1 to 148.0	144.0 to 148.0	144.1 to 148.0

APPENDIX 3
Table of Emissions

The FCC is now using the new WARC emission symbols. In the new system, the old 2-character symbols have been replaced with 3-character symbols. The 3-character symbols give more specific information concerning the emissions that they represent.

FIRST CHARACTER
- N   Emission of an unmodulated carrier
- A   AM double-sideband
- J   Single sideband, suppressed carrier
- F   Frequency modulation
- P   Sequence of unmodulated pulses
- C   Vestigial sidebands

SECOND CHARACTER
- 0   No modulating symbol
- 1   Digital information - no modulation
- 2   Digital information with modulation
- 3   Modulated with analog information

THIRD CHARACTER
- N   No information transmitted
- A   Telegraphy for reception by air
- B   Telegraphy for automatic reception
- C   Facsimile
- D   Data transmission, telemetry, telecommand
- E   Telephony
- F   Television

Traditional Symbol		New Symbol
AMPLITUDE MODULATED		
Unmodulated	A0	N0N
Keyed on/off	A1	A1A
Tones keyed on/off	A2	A2A
AM data		A2D
Keyed tones w/SSB	A2J	J2A
SSB data		J2D
AM voice	A3	A3E
Voice w/SSB	A3J	J3E
AM facsimile	A4	A3C
SSB television	A5	C3F
AM television	A5	A3F
FREQUENCY MODULATED		
Unmodulated	F0	N0N
Switched between two frequencies	F1	F1B
Switched tones	F2	F2A
FM data		F2D
FM voice	F3	F3E
FM facsimile	F4	F3C
FM television	F5	F3F
PULSE MODULATED		
Phase	P	P1B

CODE PRACTICE OSCILLATOR and MONITOR

Model OCM-2

- Large 3" speaker with volume and tone controls.
- Latest IC circuitry for high quality sound.
- Professional equipment at a very low price.
- Converts to a CW monitor. • Kit form or wired.

Model OCM-2 is a professional, high quality code practice oscillator with an attractive two-color panel. Its 3" speaker and advanced IC circuitry permit a high quality sound which can be varied in volume and pitch. Speaker or headphones can be used. Once the code has been learned, the OCM-2 can easily be converted into a CW monitor for use with the transmitter. Model OCM-2 is economically priced and is available in kit form or wired. Easy-to-follow instructions make kit-building an interesting, educational project.

Model OCM-2K - in kit form $17.50
Model OCM-2W - factory wired $22.95

AMECO EQUIPMENT DIVISION
AMECO PUBLISHING CORP.
220 East Jericho Turnpike
Mineola, New York 11501
(516) 741-5030

THE COMPLETE MORSE CODE COURSE FOR THE PC

*Generate random characters at ANY speed

*Generate random QSO's-similar to the VEC exams - at ANY speed

*Send text from any external data file

*Complete lesson on learning the Morse Code

*Includes 32 page book on Code learning and user's manual

*Plus many, many more features

Ameco's Morse Code course for the PC is the most versatile program of its kind. It is user friendly and menu driven, containing over 18 options. It will run on any IBM PC/XT/AT (or 100 percent compatible) at any clock speed, in either monochrome or color.

There are many other features, including quiz sessions for the beginner, as well as the ability to alter letter, character and word spacing to simulate HI/LO spacing. The program also will turn your keyboard into a straight or iambic keyer.

This course is ideal for the beginner, and perfect for the licensed ham who wishes to upgrade! All from AMECO Publishing, the oldest and largest publisher of code training material, for over 38 years.

AMECO'S Morse Code Course for the PC (Cat. No. 107-PC)...$19.95

TUNABLE PREAMPLIFIER ANTENNA
Model TPA

- Can be used as a tunable preamplifier or an indoor active antenna.
- Complete coverage from 0.22 to 30 MHz.
- Improves receiver gain and noise figure.
- Over 20 dB gain on all frequencies.

Model TPA is a dual function unit. It can be used as a preamplifier to improve the gain of a receiver, or as an indoor active antenna when an outdoor antenna is not available.

Model TPA contains a tuned RF amplifier that covers all frequencies from 0.22 to 30 MHz., including amateur bands, all foreign broadcast bands, citizen's band and all other services within this range. A dual gate, field effect transistor provides an excellent noise figure and over 20 dB gain. The weak signal performance of most receivers is improved.

Model TPA uses either an internal 9 volt battery or an AC adapter, such as Ameco Model P-9T. As a preamplifier, the input matches most antennas. Long wire, 300 ohm and random length antennas can also be used with good results.

When no external antenna is available, the preamplifier's whip antenna gives excellent results.

Model TPA. Preamplifier/Antenna $74.95
Model P-9T. AC adapter for TPA .... $ 7.50

Ameco Equipment Div. of Ameco Publishing Corp.
220 East Jericho Turnpike, Mineola, NY 11501
(516) 741-5030